GOOD SCIENCE

VICTORIAN CURRICULUM
YEAR 10

EMMA **CRAVEN**
AARON **ELIAS**

With contributions from:
Rebecca **Cashmere**
Haris **Harbas**
Niru **Kumar**
Natalie **Stinson**

Investigations reviewer:
Margaret **Croucher**

Good Science 10 Victorian Curriculum
1st edition

Emma Craven
Aaron Elias
Rebecca Cashmere
Haris Harbas
Niru Kumar
Natalie Stinson

Publisher: Catherine Charles-Brown
Publishing director: Olive McRae
Project editor: Sarah Blood
Proofreader: Kelly Robinson
Indexer: Max McMaster
Typesetters: Paul Ryan, Beau Lowenstern
Cover and text designer: Regine Abos
Production controller: Sarah Blake
Digital production: Erin Dowling
Permissions researcher: Hannah Tatton
Illustrator: QBS Learning
Cover illustrator: Dominic D'Monte

First published 2021 by
Matilda Education Australia, an imprint of Meanwhile Education Pty Ltd
Melbourne, Australia
T: 1300 277 235
E: customersupport@matildaed.com.au
www.matildaeducation.com.au

National Library of Australia Cataloguing-in-Publication data
Author: Emma Craven, Aaron Elias
Title: Good Science Victorian Curriculum 10
ISBN: 9780655090090

A catalogue record for this book is available from the National Library of Australia at www.nla.gov.au

Warning: It is recommended that Aboriginal and Torres Strait Islander peoples exercise caution when viewing this publication as it may contain names or images of deceased persons.

Printed in Malaysia by Vivar Printing
Sep-2023

MAKE EVERY LESSON A **GOOD** LESSON

CONTENTS

CURRICULUM CORRELATION GRID

VICTORIAN CURRICULUM SCIENCE **LEVELS 9 AND 10**		
SCIENCE UNDERSTANDING: SCIENCE AS A HUMAN ENDEAVOUR		
VCSSU114	Scientific understanding, including models and theories, are contestable and are refined over time through a process of review by the scientific community	• Chapter 3: The periodic table • Chapter 6: The universe • Chapter 7: Energy
VCSSU115	Advances in scientific understanding often rely on developments in technology and technological advances are often linked to scientific discoveries	• Chapter 1: Genetics • Chapter 6: The universe • Chapter 7: Energy
VCSSU116	The values and needs of contemporary society can influence the focus of scientific research	• Chapter 1: Genetics • Chapter 4: Chemical reactions • Chapter 5: Global systems • Chapter 7: Energy
SCIENCE UNDERSTANDING: BIOLOGICAL SCIENCES		
VCSSU117	Multicellular organisms rely on coordinated and interdependent internal systems to respond to changes to their environment	• Good Science 9
VCSSU118	An animal's response to a stimulus is coordinated by its central nervous system (brain and spinal cord); neurons transmit electrical impulses and are connected by synapses	• Good Science 9
VCSSU119	The transmission of heritable characteristics from one generation to the next involves DNA and genes	• Chapter 1: Genetics
VCSSU120	The theory of evolution by natural selection explains the diversity of living things and is supported by a range of scientific evidence	• Chapter 2: Evolution
VCSSU121	Ecosystems consist of communities of interdependent organisms and abiotic components of the environment; matter and energy flow through these systems	• Good Science 9
SCIENCE UNDERSTANDING: CHEMICAL SCIENCES		
VCSSU122	All matter is made of atoms which are composed of protons, neutrons and electrons; natural radioactivity arises from the decay of nuclei in atoms	• Good Science 9
VCSSU123	The atomic structure and properties of elements are used to organise them in the periodic table	• Chapter 3: The periodic table
VCSSU124	Chemical reactions involve rearranging atoms to form new substances; during a chemical reaction mass is not created or destroyed	• Chapter 4: Chemical reactions • Good Science 9
VCSSU125	Different types of chemical reactions are used to produce a range of products and can occur at different rates; chemical reactions may be represented by balanced chemical equations	• Chapter 4: Chemical reactions
VCSSU126	Chemical reactions, including combustion and the reactions of acids, are important in both non-living and living systems and involve energy transfer	• Good Science 9
SCIENCE UNDERSTANDING: EARTH AND SPACE SCIENCES		
VCSSU127	The theory of plate tectonics explains global patterns of geological activity and continental movement	• Good Science 9
VCSSU128	Global systems, including the carbon cycle, rely on interactions involving the atmosphere, biosphere, hydrosphere and lithosphere	• Chapter 5: Global systems
VCSSU129	The Universe contains features including galaxies, stars and solar systems; the Big Bang theory can be used to explain the origin of the Universe	• Chapter 6: The universe

SCIENCE UNDERSTANDING: PHYSICAL SCIENCES		
VCSSU130	Electric circuits can be designed for diverse purposes using different components; the operation of circuits can be explained by the concepts of voltage and current	• Good Science 9
VCSSU131	The interaction of magnets can be explained by a field model; magnets are used in the generation of electricity and the operation of motors	• Good Science 9
VCSSU132	Energy flow in Earth's atmosphere can be explained by the processes of heat transfer	• Chapter 7: Energy
VCSSU133	The description and explanation of the motion of objects involves the interaction of forces and the exchange of energy and can be described and predicted using the laws of physics	• Chapter 8: Motion
SCIENCE INQUIRY SKILLS: QUESTIONING AND PREDICTING		
VCSIS34	Formulate questions or hypotheses that can be investigated scientifically, including identification of independent, dependent and controlled variables	• Skills and investigations • Good Science 9
SCIENCE INQUIRY SKILLS: PLANNING AND CONDUCTING		
VCSIS135	Independently plan, select and use appropriate investigation types, including fieldwork and laboratory experimentation, to collect reliable data, assess risk and address ethical issues associated with these investigation types	• Skills and investigations • Good Science 9
VCSIS136	Select and use appropriate equipment and technologies to systematically collect and record accurate and reliable data, and use repeat trials to improve accuracy, precision and reliability	• Skills and investigations • Good Science 9
SCIENCE INQUIRY SKILLS: RECORDING AND PROCESSING		
VCSIS137	Construct and use a range of representations, including graphs, keys, models and formulas, to record and summarise data from students' own investigations and secondary sources, to represent qualitative and quantitative patterns or relationships, and distinguish between discrete and continuous data	• Skills and investigations • Good Science 9
SCIENCE INQUIRY SKILLS: ANALYSING AND EVALUATING		
VCSIS138	Analyse patterns and trends in data, including describing relationships between variables, identifying inconsistencies in data and sources of uncertainty, and drawing conclusions that are consistent with evidence	• Skills and investigations • Good Science 9
VCSIS139	Use knowledge of scientific concepts to evaluate investigation conclusions, including assessing the approaches used to solve problems, critically analysing the validity of information obtained from primary and secondary sources, suggesting possible alternative explanations and describing specific ways to improve the quality of data	• Skills and investigations • Good Science 9
SCIENCE INQUIRY SKILLS: COMMUNICATING		
VCSIS140	Communicate scientific ideas and information for a particular purpose, including constructing evidence-based arguments and using appropriate scientific language, conventions and representations	• Skills and investigations • Good Science 9

GENETICS

What makes us who we are?

The greatest leaps forward in science might be found in the smallest of places. In nearly every cell of your body, there is a tiny molecular blueprint called DNA. Scientific advances mean that DNA mapping that once took 15 years and cost $5 billion can now be done rapidly for less than $100. We can find out what diseases we are most at risk of, what medicines work best for us individually and what cultural heritage we may have but otherwise would never have known about. These advancements in biotechnology bring with them the potential for both miracles such as cell regeneration and terrors such as biological warfare.

1 FRAYER MODEL

Copy and complete the below chart in your workbook.

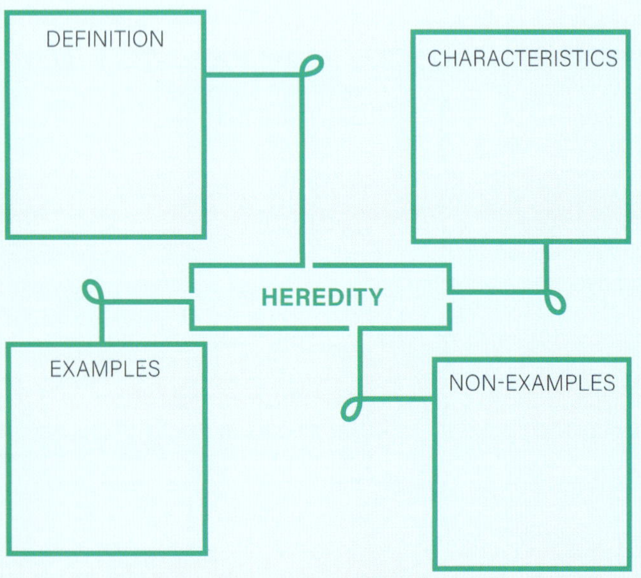

Complete two additional charts for the key terms *DNA* and *Phenotype*.

2 LEARNING LINKS

Brainstorm everything you already know about genetics.

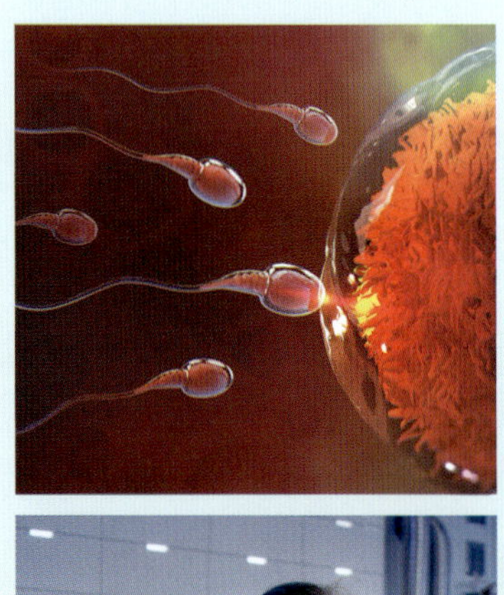

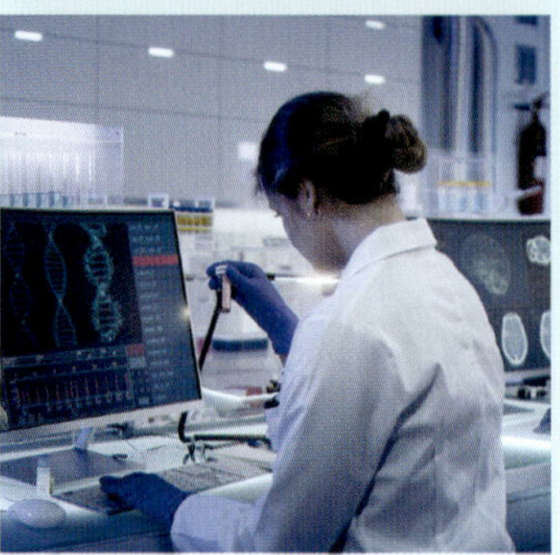

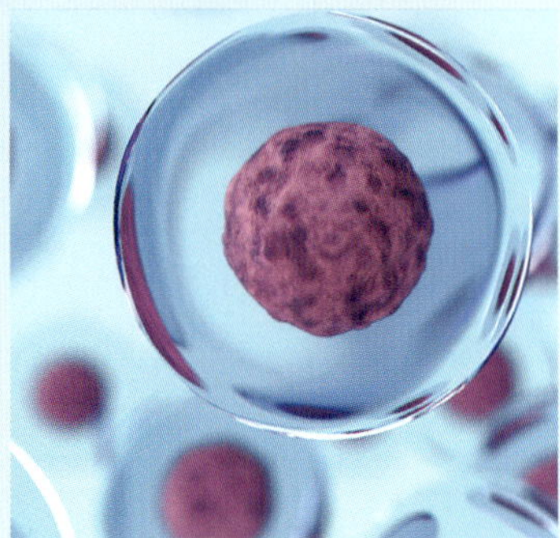

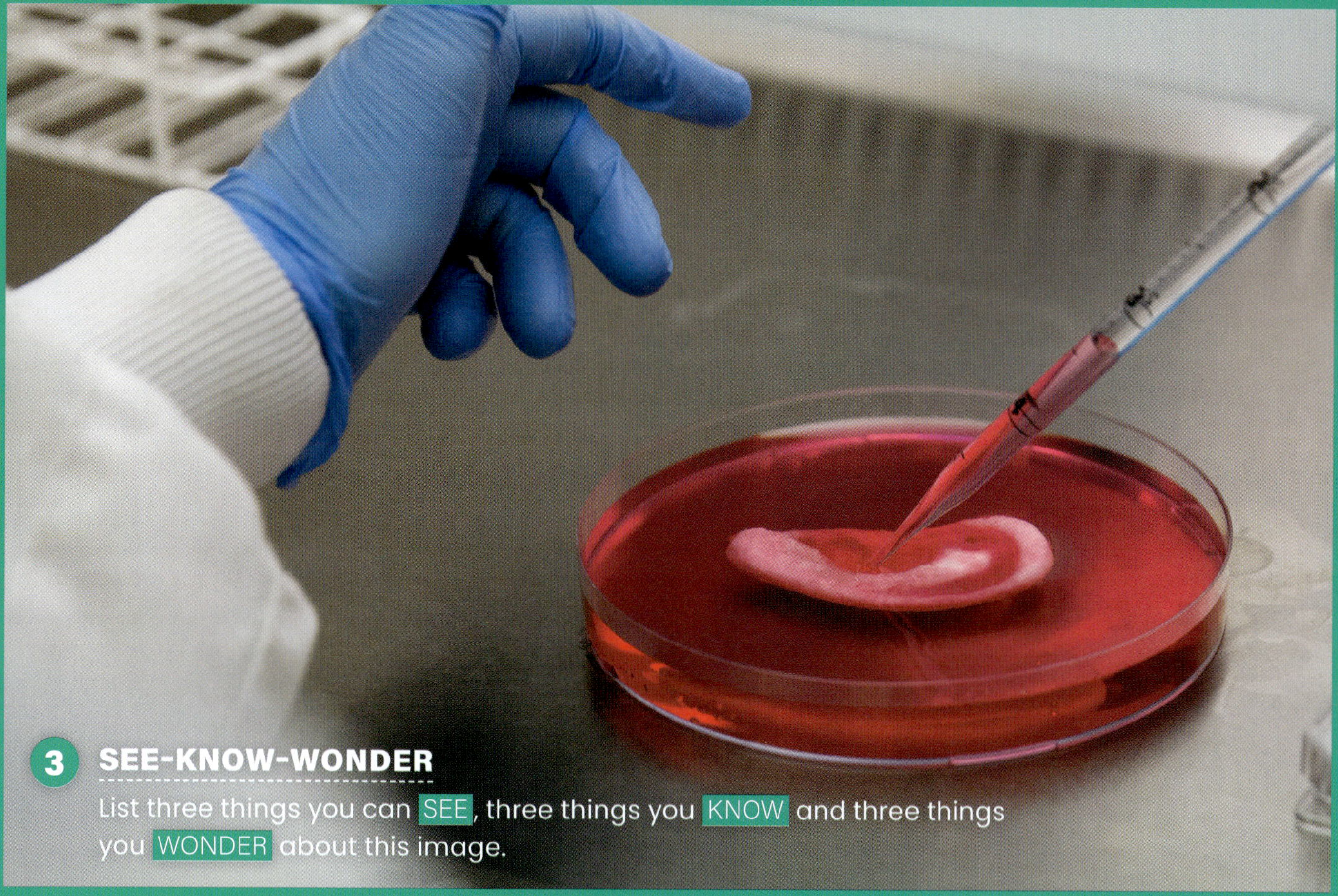

3 SEE-KNOW-WONDER

List three things you can SEE, three things you KNOW and three things you WONDER about this image.

4 CRITICAL + CREATIVE THINKING

FIVE QUESTIONS: Write five questions that have the answer 'DNA'.

WHAT IF ... parents could choose the characteristics of their baby before it was born?

THE BAR: Think of one thing you would make **B**igger, one thing you would **A**dd and one thing you would **R**eplace on a human being.

5 THE BIGGEST PROJECT!

In 1990, the Human Genome Project began. This was an international research project with the goal of mapping the entire human genome (all of human DNA) of more than 3 billion base codes. Scientists from all over the world worked for 15 years on research estimated to cost more than $5 billion. Because of the wonderful work on the Human Genome Project, scientists were able to identify and locate all the genes in human DNA.

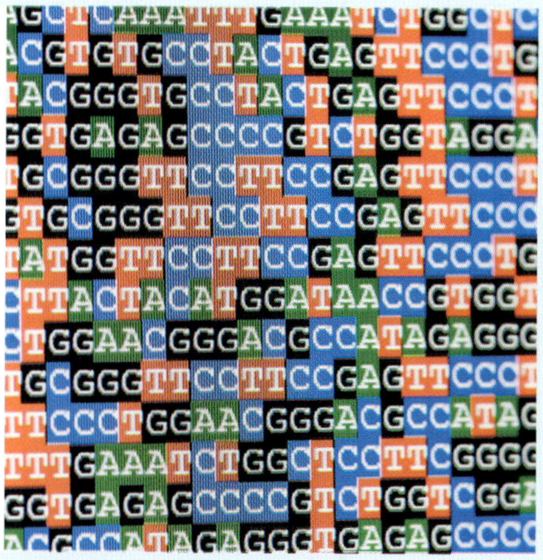

1.1 HEREDITY

LEARNING INTENTION

At the end of this lesson I will be able to describe how genetic information is passed from parents to offspring.

KEY TERMS

chromosome
a tightly coiled strand of DNA

fertilisation
the fusion of male and female gametes to form a zygote

gamete
a sex cell – an ovum or a sperm

gene
a segment of DNA, the basic functional unit of heredity

heredity
the passing of traits from parents to their offspring

meiosis
complex cell division that produces gametes (sex cells)

ovum
a female sex cell, also known as an egg

predisposition
the increased likelihood of developing a disease or condition

sperm
a male sex cell

NUMERACY LINK

CALCULATION

The human egg is about 0.17 mm in diameter and a sperm cell is about 0.05 mm in length. How many times bigger is the egg cell than a sperm cell?

The genetic information used to make a unique and incredible life (that's you!) comes from both your mother and your father. Men produce about 1500 **sperm** every second, while women release an egg every month. To make a baby, these sperm must survive and traverse the female reproductive system. Only one of the sperm will **fertilise** an egg to begin the process of creating new life.

1 Heredity is the inheritance of traits

Heredity is the passing of traits or characteristics from parents to their offspring. Traits that can be genetically inherited from parents are things such as eye colour, straight or curly hair and even the ability to roll your tongue. Unfortunately, less desirable traits can be passed down too – including **predispositions** to certain diseases like cancer. Predisposition does not mean you will get that disease, just that you are more likely to than someone without the predisposition. In humans (and many other animals), heredity occurs through sexual reproduction. Half of your genetic information comes from your mother's **ovum** (egg) and half from your father's sperm.

What kinds of traits can be inherited genetically?

2 Certain characteristics are passed down from parents

Genetic information is passed down from parents to offspring in organised little packages called **chromosomes**. Human cells typically have 46 chromosomes; 23 are inherited from the mother and 23 from the father. During fertilisation, when the 23 chromosomes from the ovum fuse with the 23 from the sperm, the zygote (first cell) has a complete set of 46 chromosomes. These chromosomes contain all the **genes** the individual will ever have: a biological instructional manual for how to make a unique person.

This combination of genes from your mother and father determines your characteristics. Some traits, such as height, involve many genes.

How are some characteristics passed down from parents to offspring?

Figure 1.1 A sperm penetrates an ovum (egg) to produce a new cell.

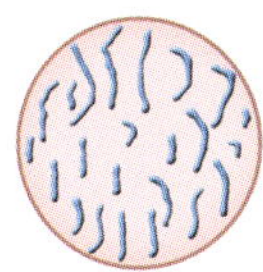

Mother
Ovum (egg)
23 single chromosomes

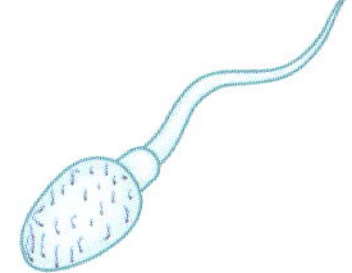

Father
Sperm 23 single chromosomes

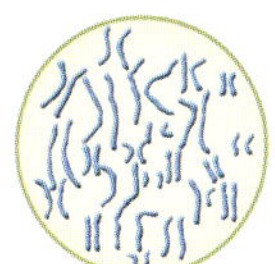

Fertilised egg
46 paired chromosomes

Figure 1.2 The sperm and ovum contain 23 chromosomes each, which make a full set of 46 chromosomes in the fertilised egg.

❸ Meiosis is how sex cells are created

Human reproductive cells are called **gametes**. In males, the gamete is the sperm, while in females the gamete is the ovum (egg). These cells are not created like normal body cells using mitosis; they are formed in a special process called **meiosis**.

First, the cell divides in half to create two new daughter cells, as happens during mitosis. Unlike in mitosis, each of these cells divides again so there are four daughter cells. Each is different – to both its parent *and* its sister.

When a parent cell divides for the first time during meiosis, the chromosome threads don't split in half. Instead, the nucleus splits in half, and each half holds half of the threads. When the rest of the cell divides, each half carries one of the nucleus halves, which then become the nuclei of the new cells. This means that each daughter cell only carries half of the chromosomes of the parent cell.

During reproduction, when the sperm and egg unite to form a single cell, the number of chromosomes is restored in the offspring. The fertilised egg has all the chromosomes needed to be a functioning cell.

In what type of cell does meiosis happen?

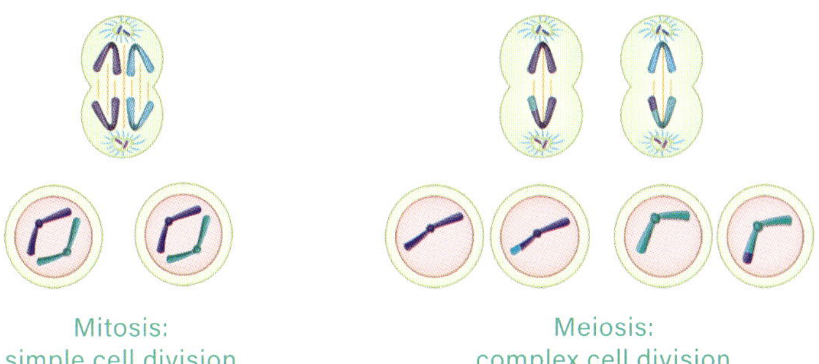

Figure 1.3 The main difference between mitosis (left) and meiosis (right) is how much of the parent cell's genetic information is transferred to the daughter cells.

INVESTIGATION 1.1
Genetic trait survey

KEY SKILL
Using primary and secondary scientific evidence

▶ *Go to page **136***

CHECKPOINT 1.1

1 What are the male and female gametes (sex cells) called?
2 How many chromosomes should human (non-sex) cells possess?
3 How many chromosomes should gametes (sex cells) possess?
4 Give five examples of characteristics that can be inherited genetically.
5 Describe the process of fertilisation.
6 Explain meiosis in your own words.
7 Suggest why it is important that gametes have only half the genetic information compared to non-sex cells.

EXTENSION

8 Compare and contrast meiosis with mitosis. How are they similar and how are they different?

SUCCESS CRITERIA

- I can describe how hereditary characteristics are passed from parents to offspring.
- I can define and describe the processes of fertilisation and meiosis.

1.2 DNA, GENES AND CHROMOSOMES

LEARNING INTENTION

At the end of this lesson I will be able to describe the relationship between DNA, genes and chromosomes.

KEY TERMS

deoxyribonucleic acid (DNA)
found in the nucleus of a cell; the carrier of genetic information

double helix
the structure of a DNA molecule; a double-stranded spiral

genome
an organism's entire set of DNA

genotype
the genetic code for a gene or an organism's entire genome

nucleotide
building block of DNA (adenine, guanine, thymine or cytosine)

phenotype
how a genotype is physically expressed

LITERACY LINK

VOCABULARY

Use the key terms in this lesson as well as three from the previous lesson to play vocabulary bingo. Draw up a grid of nine squares and write one key term in each square at random. Your teacher will read definitions and, if you have the corresponding term, you can strike it off your bingo card.

Your DNA is a blueprint completely unique to you. It contains all the instructions for your genetic traits, passed down from your mother and father. These characteristics are passed down on genes – sections of DNA. Some genes are common (such as those for brown eyes) and some are rarer (such as those for red hair). Genes are passed down on chromosomes; humans typically have 46 chromosomes.

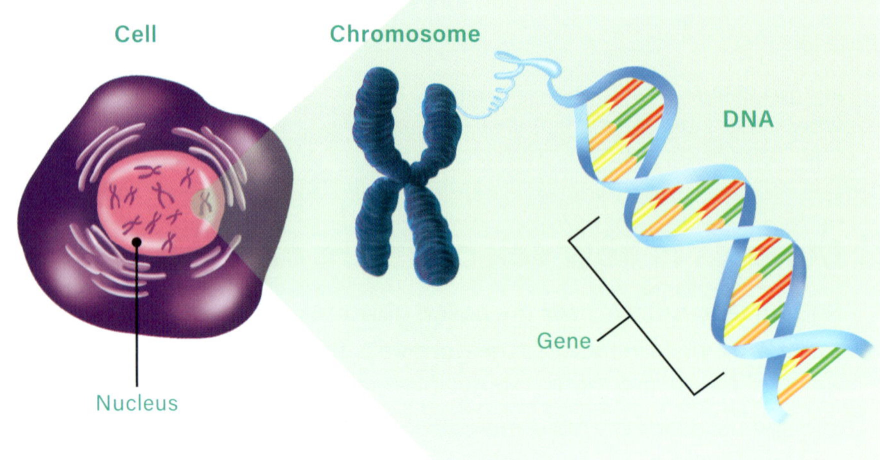

Figure 1.4 Chromosomes contain DNA, which in turn is made up of genes.

❶ DNA is made up of nucleotides

Deoxyribonucleic acid (**DNA**) is the blueprint that contains all our biological instructions. Found in the nucleus of nearly every cell, DNA is in the shape of a twisted ladder called a **double helix**. The sides of the 'ladder' are the backbone of DNA and are made up of units called **nucleotides**.

Each nucleotide is made up of a phosphate group, a five-carbon sugar and a nitrogenous base. There are four different nitrogenous bases – adenine, thymine, guanine and cytosine – which pair together like the rungs of a ladder. The bases only partner in specific pairs – adenine pairs with thymine and guanine pairs with cytosine.

Where is DNA found?

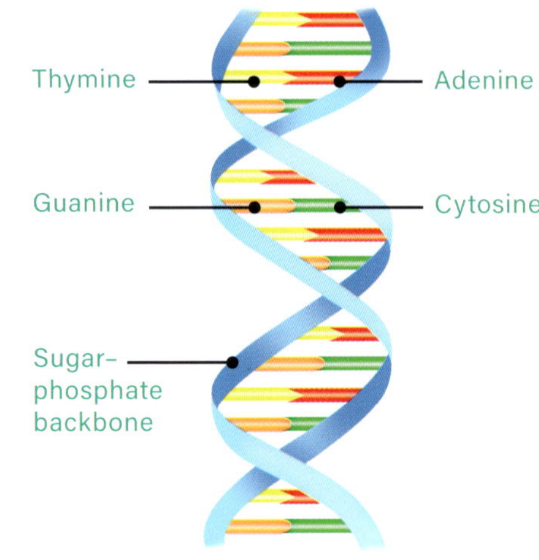

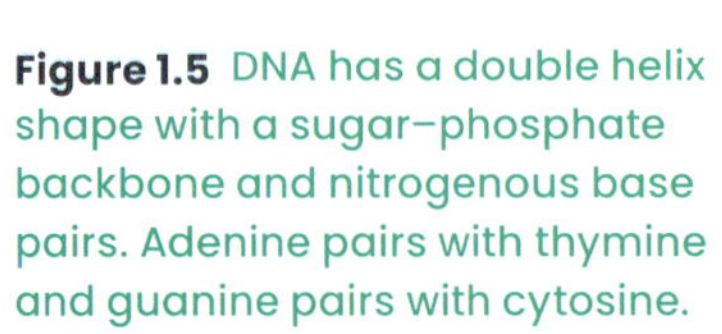

Figure 1.5 DNA has a double helix shape with a sugar–phosphate backbone and nitrogenous base pairs. Adenine pairs with thymine and guanine pairs with cytosine.

❷ Genes are segments of DNA

One section of DNA is called a gene. A gene is the basic functional unit of heredity. Genes are how traits are passed down from mother and father to their offspring. Every human has two copies of each gene; one from their mother and one from their father.

Some genes code for proteins. This is very important because proteins:

- are used to repair and make new tissue
- make up much of our skin, hair, muscle and bone
- make up the chemical messenger system – hormones, blood and antibodies.

Other genes code for a variety of molecules with different functions. Genes can vary in size – they can be a few hundred bases long or millions of bases long. Your genes make up your **genotype** – the genetic code for a specific gene or genes – and determine your unique genetic identity (your **genome**). Your genotype codes for your **phenotype** – how those genes will be expressed. This can be through visible traits, such as hair or eye colour, or characteristics that aren't visible, such as blood type.

Figure 1.6 A gene is a segment of DNA.

Gene

What is a gene?

❸ Condensed chromosomes contain tightly coiled DNA

DNA often exists in cells as disorganised long pieces. Under a microscope, DNA can look like a bunch of squiggles. At various stages (especially during cell division) the DNA in chromosomes becomes tightly coiled and packed.

Each chromosome is made up of two identical chromatids. The two chromatids are joined so that they have short and long arms. Each chromatid contains one long DNA molecule that has been tightly coiled and packed into the small structure. In order to ensure the DNA is effectively coiled and packed, it is wrapped around little protein balls called histones.

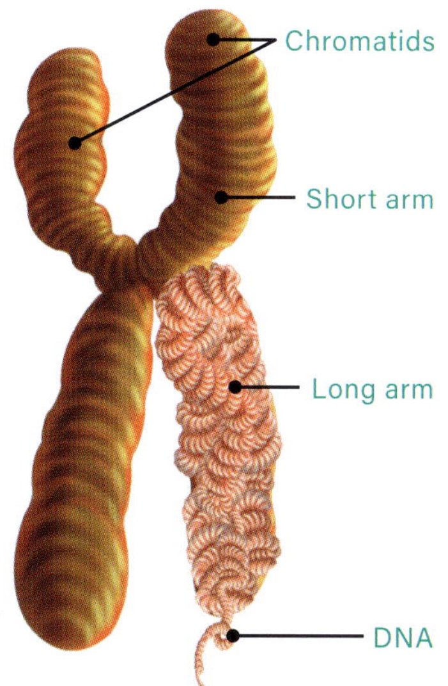

Figure 1.7 At some stages of the cell cycle, chromosomes are tightly coiled strands of DNA.

What is a chromosome made of?

INVESTIGATION 1.2A

Extracting DNA from strawberries

KEY SKILL
Identifying and managing relevant risks

▶ *Go to page **138***

INVESTIGATION 1.2B

DNA lolly model

KEY SKILL
Using modelling and simulations

▶ *Go to page **139***

CHECKPOINT 1.2

1 What does DNA stand for?

2 Explain the differences between DNA and genes.

3 Which DNA base pairs with adenine?

4 What is the backbone of DNA made up of?

5 What is the difference between phenotype and genotype?

6 Copy and complete.
  a A ________ is a section of DNA.
  b DNA is shaped like ________
  c Your genome is your ________

CRITICAL AND CREATIVE THINKING

7 Design a model that represents the relationship between DNA, genes and chromosomes.

SUCCESS CRITERIA

- I can explain the relationship between DNA and genes.
- I can describe and draw the structure of DNA.

1.3 PATTERNS OF INHERITANCE

LEARNING INTENTION

At the end of this lesson I will be able to investigate the patterns of inheritance of characteristics through generations of a family.

KEY TERMS

allele
a different version of a gene

heterozygote
a gene that has two different alleles present

homozygote
a gene that has identical alleles present

LITERACY LINK

READING

Read through this lesson and highlight any words you are unfamiliar with. Spend five minutes going over the definitions of these words then explain them in your own words to a peer.

NUMERACY LINK

UNITS

Change the following numbers to Roman numerals: 2, 5 and 10.

Are you in a family where everyone's eye or hair colour is different? Have you ever wondered why red hair or green eyes are so rare? Some traits are less likely to occur than others, and there are mathematical ways to figure out what traits are likely to be passed down from parent to offspring.

1 Alleles can be dominant or recessive

Each of your 23 pairs of chromosomes contains one from your father and one from your mother. This means you end up with two versions of each gene, and these different versions are called **alleles**. For example, you might get a blue eye-colour allele from your mother and a brown eye-colour allele from your father. But how can you figure out what your eye colour will be? Well, some alleles are dominant while others are recessive. Where a dominant allele is paired with a recessive allele, the dominant trait will always be expressed.

If the trait of brown eyes is dominant and the trait of blue eyes is recessive, then even if both alleles are present, the brown will be dominant over the blue and the offspring will have brown eyes. Blue-eyed offspring will only occur if both alleles present are for blue eyes. The genotype for a dominant allele is an upper-case letter and for a recessive allele a lower-case letter.

In what situation will a recessive trait be expressed?

2 Punnett squares can predict the genotypes of offspring

Continuing with the eye-colour example, it is possible to use a diagram called a Punnett square to predict the genotypes of offspring based on the genotypes of their parents. First, you identify whether the parents' genotypes are **homozygous** or **heterozygous**. A homozygote is when both of an individual's alleles are the same; for example, BB or bb. A heterozygote is when the two alleles are different; for example, Bb.

To draw a Punnett square, you draw up a grid like the ones in Figure 1.9, with the genotypes of one parent across the top and of the other parent down the left side. You then complete the table by combining the genotypes from both parents into the corresponding squares. If both parents have the heterozygous genotype Bb, there is a 75% chance of their offspring inheriting brown eyes and only a 25% chance of blue eyes.

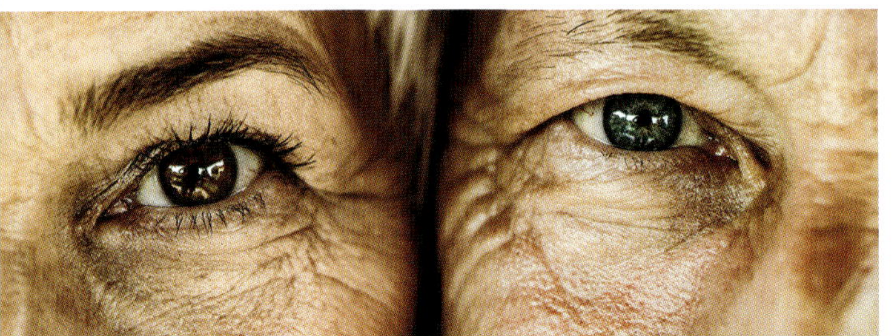

Figure 1.8 Blue eyes only occur when the alleles inherited from both parents are for blue eyes.

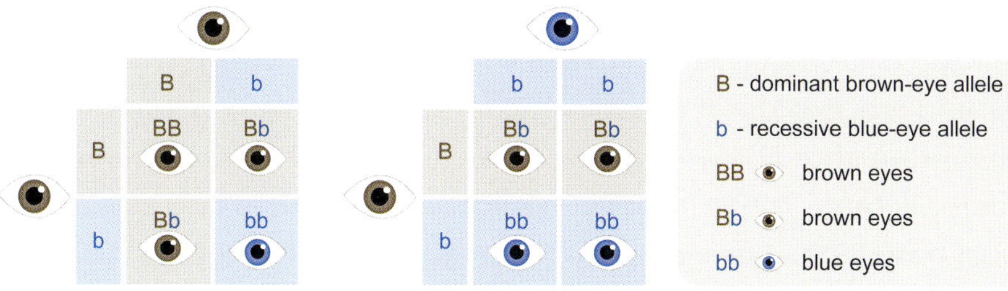

Figure 1.9
Punnett squares show the likelihood of inheriting certain genetic traits.

What is the difference between heterozygotes and homozygotes?

❸ Pedigree charts show traits passed down in families

A pedigree chart is a diagram that shows which family members possess a particular trait or phenotype over generations, and can be used to identify whether a trait is dominant or recessive. Males are represented by a square and females by a circle. If there is a line between them it means they are a mating pair, and if they have children a line will come down from the middle of their mating line. If a square or circle is coloured in, it means they possess the trait. A pedigree chart is labelled on the left in Roman numerals to number each generation.

In this famous pedigree chart you can see the occurrence of haemophilia (a rare blood disease) in the descendants of Queen Victoria. The gene for haemophilia is recessive and also linked to sex. As you can see in the pedigree chart, only males had this disease (though females carried the gene).

What symbol indicates a female possessing a genetic trait?

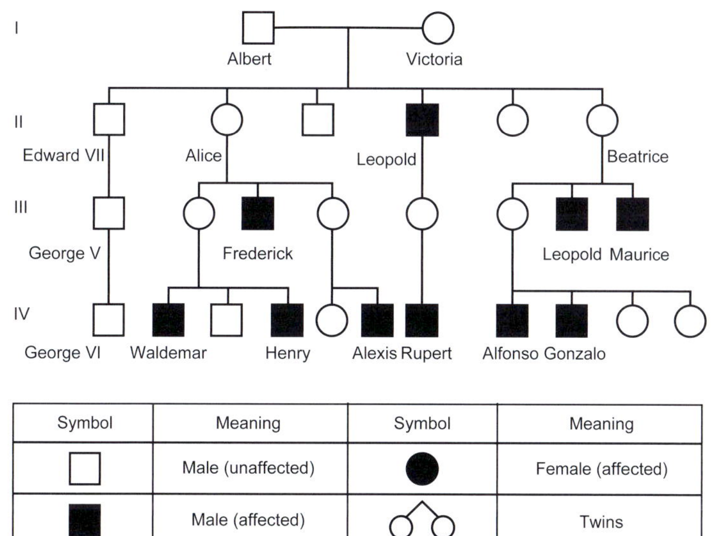

Symbol	Meaning	Symbol	Meaning
□	Male (unaffected)	●	Female (affected)
■	Male (affected)	twins symbol	Twins
○	Female (unaffected)	□—○	Mating

Figure 1.10
Many of the male descendants of Queen Victoria suffered from haemophilia.

CHECKPOINT 1.3

1 Describe the uses of a Punnett square and a pedigree chart.

2 What does a coloured-in circle represent in a pedigree chart?

3 Explain why everyone has two copies of each allele.

4 Create a Punnett square that crosses two parents with the genotypes Aa and aa.
   a What percentage of the offspring will be homozygotes and what percentage will be heterozygotes?
   b Is the genotype aa an example of a recessive or dominant homozygote?

5 Explain the following statement.
   Characteristics like hair colour can be dominant or recessive.

6 Give an example of a homozygous and a heterozygous genotype.

7 Can you tell from an organism's phenotype what its genotype is? Explain your answer.

CONNECTING IDEAS

8 Create a pedigree chart showing your family, beginning with your grandparents' generation. Ensure you are using correct symbols. If you can, add one trait (for example, eye colour or hair colour).

SUCCESS CRITERIA

- I can complete a simple Punnett square.
- I can read and understand a simple pedigree chart.
- I can describe alleles in terms of dominance and recessiveness.

1.4 GENETIC ABNORMALITIES

LEARNING INTENTION

At the end of this lesson I will be able to describe mutations as changes in DNA or chromosomes, and outline the factors that contribute to causing mutations.

KEY TERMS

karyotype
a picture of an organism's full set of chromosomes

mutation
an error in DNA caused by a substitution, an insertion or a deletion of a base

ribonucleic acid (RNA)
a nucleic acid that carries instructions from DNA

LITERACY LINK

WRITING

Write a numbered step-by-step process outlining how DNA is replicated.

NUMERACY LINK

CALCULATION

Three bases (a base triplet) on a strand of DNA codes for an amino acid. How many different ways can you arrange a base triplet using the 4 bases?

Let's consider how large and intricate DNA is. If you set each strand of DNA from every cell in your body end to end, they would reach around the solar system, twice! It makes sense, therefore, that sometimes mistakes happen when DNA is copied or replicated. These errors can become mutations and can affect just one gene, many genes or even entire chromosomes.

In order to understand where errors can occur in either DNA or chromosomes, we first need to explore how DNA is replicated.

1 Errors can occur in DNA replication

DNA is replicated whenever a new cell is formed. In this process, one strand of old DNA acts as a template from which a new strand is produced. The new DNA molecule is made of one old strand and one new strand.

The old double-stranded DNA molecule is 'unzipped' by an enzyme called DNA helicase and prepared for replication by a molecule called **ribonucleic acid** (**RNA**). Once prepared by RNA, the nitrogenous bases are ready to act as a template for a strand of complementary new DNA. The new DNA strand matches up along the old strand, so thymine is only paired with adenine and guanine is only paired with cytosine.

The new DNA grows one base at a time in a process called elongation. More enzymes then 'proofread' the new strand to find any mistakes, and then again to finish and complete the process. The new DNA molecule is referred to as a 'daughter molecule'.

How does DNA replicate?

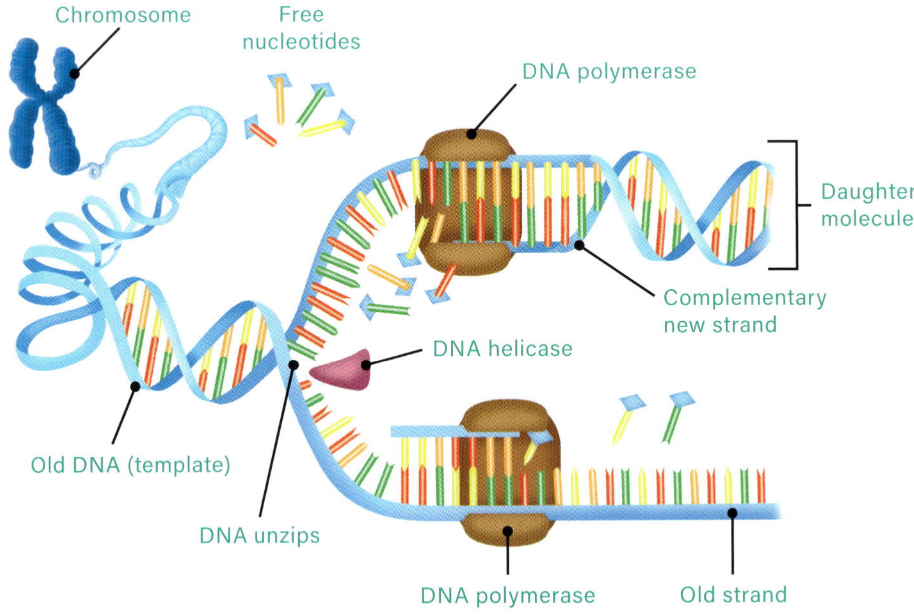

Figure 1.11 In DNA replication, an old strand of DNA acts as a template to make a new complementary strand.

❷ DNA mutations are errors that happen during replication

Even though the DNA replication process is very tightly controlled, mistakes can happen when the bases are being read or synthesised. These mistakes result in changes in the DNA being produced and are known as **mutations**.

There are three main types of mutation – substitution, insertion and deletion. In substitution, a different base is incorrectly inserted; for example, thymine is placed where cytosine should go. In insertion, one or more extra bases are inserted somewhere by mistake. In deletion, one or more bases are removed.

Mutations in DNA can also be caused by environmental factors – some examples include UV light, radiation and cigarette smoke, all of which can lead to mutations that could go on to become cancers.

Genetic mutations can cause cell death or disease to organisms, but many are not harmful. In some cases, mutations can even be beneficial.

What are mutations?

❸ Karyotypes can detect some chromosomal abnormalities

Karyotypes are pictures of an individual's chromosomes. In order to get this picture, cells containing condensed chromosomes are stained and examined under a microscope. An image of the chromosomes is taken and each chromosome is cut out, placed next to its pair and numbered. The largest chromosome is chromosome 1. By observing the appearance and number of chromosomes, scientists can check for chromosomal abnormalities.

In some cases, the number of chromosomes is different from the usual number. This is called *aneuploidy*. If a person has an additional chromosome number 21, then they have a genetic disorder called Down syndrome. Other differences that can occur include instances of more than two sex chromosomes, only one sex chromosome or additional chromosomes 13 or 18.

What is a karyotype?

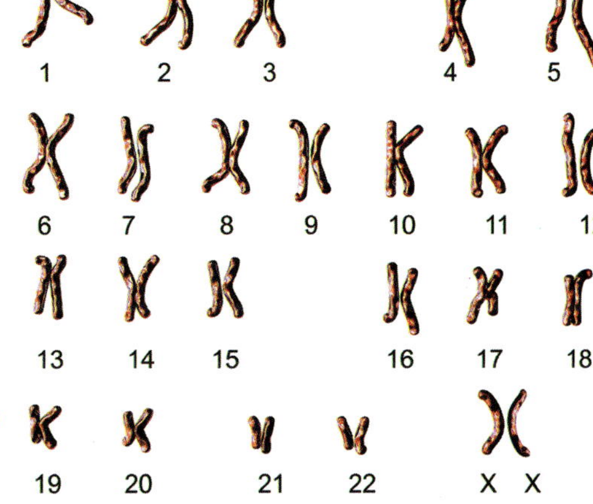

Figure 1.12 The final XX chromosome pair in this karyotype shows that the person is female. Males have an XY chromosome pair instead.

CHECKPOINT 1.4

1 Copy and complete.
DNA is replicated whenever a new _______ is formed. To replicate, one strand of _______ acts a template from which a new _______ strand is produced. This results in one _______ strand of DNA and one _______ strand.

2 Give the complementary DNA bases for the following DNA sequence: GCCAGCTTACGT.

3 Explain what a karyotype is and how it is used.

4 List four environmental factors that can lead to a mutation in DNA.

5 What are the three types of mutation?

6 What role does RNA play in DNA replication?

7 Support or reject this statement, justifying your choice.
Enzymes are not important in DNA replication.

ETHICAL CAPABILITY

8 Technology will soon allow for more comprehensive genetic screening and testing of babies before they are born. What are some ethical considerations in relation to genetic screening?

SUCCESS CRITERIA

- I can describe mutations in both DNA and chromosomes.
- I can outline some factors that contribute to causing mutations.

1.5 DEVELOPMENTS IN BIOTECHNOLOGY

LEARNING INTENTION

At the end of this lesson I will be able to describe, using examples, how the values and needs of contemporary society can influence the focus of scientific research.

KEY TERMS

biotechnology
the use of living organisms to develop products

in vitro fertilisation (IVF)
fertilisation of the sperm and egg that occurs outside the human body

stem cell
a cell type that is unspecialised and can develop into other types of cells

surrogacy
an arrangement in which a woman becomes pregnant on behalf of another person

LITERACY LINK

WRITING

Write an informative fact sheet for couples considering IVF. Include what to expect as well as any advantages or disadvantages.

NUMERACY LINK

CALCULATION

Testicles are able to make about 1500 sperm cells per second. Calculate how many sperm cells are produced per minute.

Recent developments in science and medicine have led to some incredible new technologies. However, many of these developments have been accompanied by controversy.

Figure 1.13 Dolly the sheep was the first cloned mammal and had three 'mothers'.

1 In vitro fertilisation takes place outside the body

In vitro fertilisation (**IVF**) is a type of assisted reproductive technology in which a female ovum is fertilised by a sperm outside the human body. The ova are harvested from the woman when she ovulates. Then, in a laboratory, the ova are immersed in a liquid with the sperm for fertilisation. After fertilisation, the mixture is left for a few days to allow the embryo to develop, and it is then implanted into the wall of the uterus.

IVF was first successfully completed in 1978. The process is expensive and invasive, so it is often a last resort for couples who cannot conceive naturally. IVF can also be used for **surrogacy** – in which a woman becomes pregnant and gives the baby to another person, sometimes in exchange for financial reward. The embryo is implanted into the woman (who is not the biological mother) and she carries the baby to term.

How does IVF work?

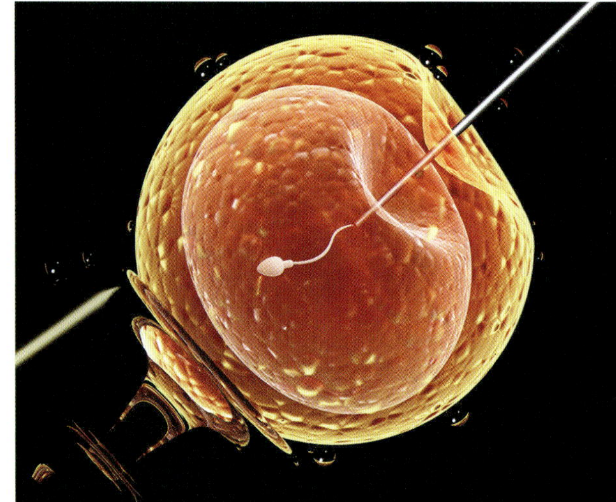

Figure 1.14 In IVF, the sperm and the egg join outside of the human body.

❷ Stem cells can develop into different types of cells

Stem cells are cells that have the potential to become many different types of specialised cells and can keep dividing to produce more cells. There are two main types of stem cells – adult stem cells (from sources such as bone marrow and intestines) and embryonic stem cells (from embryos). Adult stem cells can only develop into limited types of cells and cannot keep dividing indefinitely. Embryonic stem cells can develop into nearly any type of cell, which provides almost limitless possibilities for new treatments in medicine. Imagine being able to create new nerve cells for someone with a spinal cord injury, to grow a new heart for someone waiting for a transplant, or to repair eye tissue so someone without vision can see again. Stem cell science is still unfolding. Currently, stem cells are only used in bone marrow transplants, but their potential is exciting.

What are stem cells?

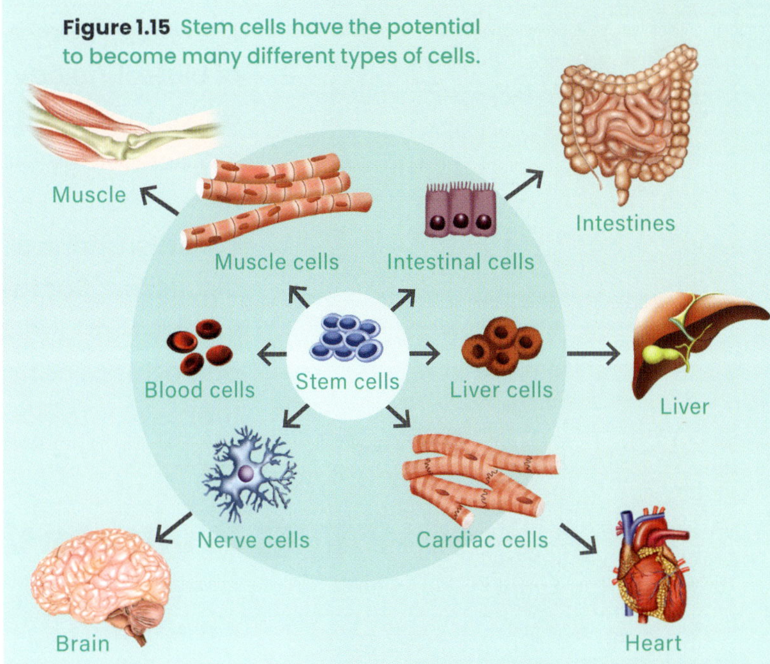

Figure 1.15 Stem cells have the potential to become many different types of cells.

❸ Biotechnology produces products from living things

Biotechnology is any use of living things to develop products or change existing ones. Biotechnology has been used for hundreds (if not thousands) of years in industries such as farming and medicine. Farmers have cultivated plant species by selecting traits over time. For example, carrots used to be purple, but after many years of selective breeding, now most carrots are orange.

Newer examples of biotechnology include genetic engineering (alterations to DNA) and cloning. Cloning is the process of producing genetically identical organisms. Vaccines are a type of biotechnology that use a dead or weak variety of microorganism to make a person immune to them. Biotechnology is used to produce antibiotics, to genetically modify food, to increase crop production and to develop crops that are resistant to pests. Biological warfare is a negative development that has resulted from developments in biotechnology.

Biotechnology has also led to the development of genetic testing and screening, where individuals can have their entire genome mapped. Genome mapping often reveals a person's cultural heritage and risks of certain health problems.

What are some examples of biotechnology?

CHECKPOINT 1.5

1 Give three examples of how developments in technology have advanced biological understanding.

2 Compare IVF to 'natural' pregnancy. What are the similarities and differences?

3 What is the difference between adult and embryonic stem cells?

4 The use of genetically modified organisms is controversial. Suggest why this might be.

5 Explain how biotechnology can be viewed as both a new and a very old science.

RESEARCH

6 Research another example of biotechnology and create a pamphlet that could be used to advertise it.

SUCCESS CRITERIA

- I can explain what biotechnology is.
- I can give at least three examples of biotechnology.

1.6 SOCIAL AND ETHICAL CONSIDERATIONS OF BIOTECHNOLOGY

LEARNING INTENTION

At the end of this lesson I will be able to discuss some advantages and disadvantages of the use and applications of biotechnology, including social and ethical considerations.

KEY TERMS

ethical consideration
a factor that takes into account what is right and what is wrong

genetically modified (GM)
an organism that has had alterations to its DNA

social consideration
a factor that affects society and the people within society

LITERACY LINK

LISTENING

Listen to a podcast about any type of IVF, genetic modification or the use of stem cells. Summarise the main points discussed in the podcast and use these to contribute to a class discussion.

There are many applications of biotechnology. Each has advantages and disadvantages. Biotechnology also has a huge impact on society and so the social and ethical considerations of each use must be discussed. **Social considerations** are factors that affect society and people and **ethical considerations** are factors that take into account what is right and what is wrong.

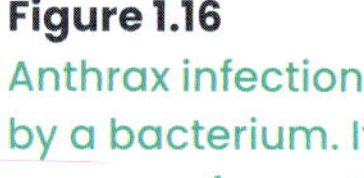

Figure 1.16
Anthrax infection is caused by a bacterium. It causes a range of symptoms, including blisters, ulcers, nausea and fever. Anthrax has been used as a biological weapon.

1 In vitro fertilisation (IVF)

Advantages

Some people are unable to conceive a child naturally because of their age, health or other factors. IVF allows them to have children.

Disadvantages

IVF is extremely expensive. Each round of treatment costs thousands of dollars. It is also invasive; the couple must undergo a number of procedures and the woman requires regular injections of hormones.

Social considerations

Not everyone can afford IVF, so it is not an equitable treatment. Should some people miss out on having a baby just because they can't afford IVF?

Ethical considerations

It has been reported that some IVF doctors have been pressured to achieve high success rates and implant several fertilised eggs into a woman, instead of just one. This has resulted in multiple births, which is less safe for the mother and the babies.

What are some advantages of IVF?

Figure 1.17
When multiple embryos are implanted, IVF has a much higher chance of multiple births. In early 2009, American Nadya Suleman gave birth to octuplets, after having had 12 embryos from IVF treatments implanted into her uterus.

❷ Stem cells

Advantages

Stem cells may be able to cure many different medical diseases and disorders, from diabetes to Parkinson's disease.

Disadvantages

Stem cell science is still new, so currently stem cells are only used fully and completely in bone marrow transplants.

What are some disadvantages of using stem cells?

Social considerations

Stem cell research is a new science, so the long-term impacts for people and society are unknown.

Ethical considerations

Harvesting embryonic stem cells destroys early-stage cells (pre-embryo). For some people who believe that life starts at conception, this practice may be considered unethical.

❸ Genetically modified crops

Advantages

Crops can be **genetically modified** (**GM**) to contain more nutrients, take up less farming land and use less water, pesticides, fertilisers and soil. GM crops also tolerate herbicides, so farmers can spray and remove weeds without damaging the crop.

Disadvantages

GM crops are pollinated just like any other crop. Some farmers are concerned that pollen from GM crops will be carried over into their non-GM crop and cross-pollinate the crop.

Social considerations

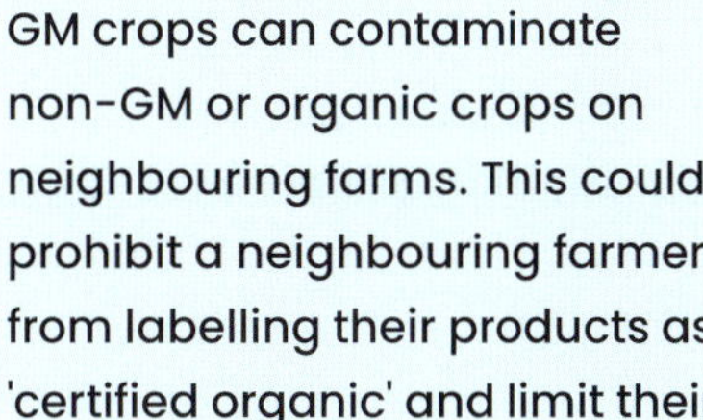

GM crops can contaminate non-GM or organic crops on neighbouring farms. This could prohibit a neighbouring farmer from labelling their products as 'certified organic' and limit their ability to earn an income.

Ethical considerations

Most core foods are now genetically modified in some way but are not labelled as GM. This means that consumers cannot make an informed choice about buying and consuming them.

What are some social and ethical considerations of genetically modified crops?

Figure 1.18 Stem cell therapies are a potential cure for diseases such as Parkinson's disease.

CHECKPOINT 1.6

1 Explain the difference between social and ethical considerations.

2 Describe some of the advantages of stem cell science.

3 Outline some of the ethical concerns about the use of embryonic stem cells and then present your own opinion.

4 A social consideration of IVF is equity. Explain why.

5 Consider biological weapons (such as anthrax). Outline some social and ethical considerations about their use.

CONNECTING IDEAS

6 Choose another type of biotechnology that is not featured in this lesson. Summarise the purpose of the biotechnology and list information under the following headings: advantages, disadvantages, social considerations and ethical considerations.

SUCCESS CRITERIA

- I can give examples of the advantages and disadvantages of biotechnology.
- I can describe some social and ethical considerations of biotechnology.

VISUAL SUMMARY

Punnett squares show the likelihood of inheriting certain genetic traits.

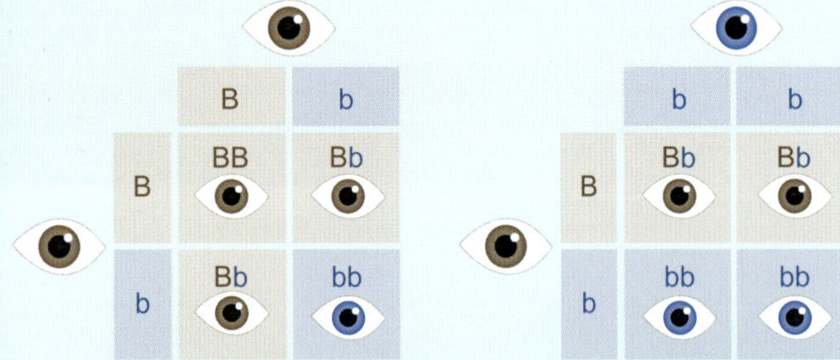

Karyotypes are pictures of an individual's chromosomes.

1 2 3 4 5
6 7 8 9 10 11 12
13 14 15 16 17 18
19 20 21 22 X X

DNA is the blueprint that contains all our biological instructions. A **gene** is a section of DNA.

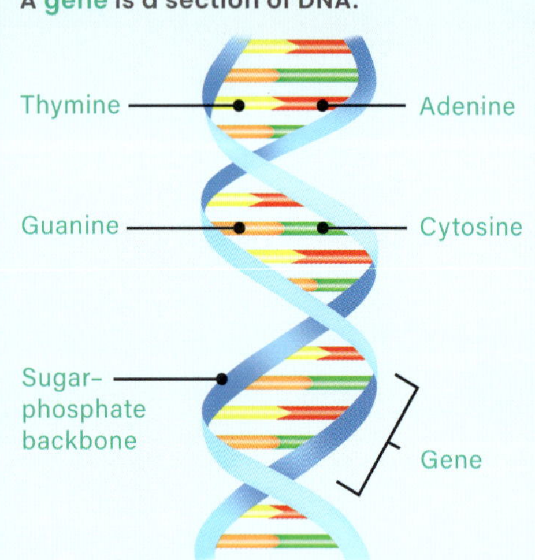

Genes are passed down in **chromosomes**.

A fertilised human egg has 46 paired chromosomes.

There are many social and ethical considerations of biotechnology.

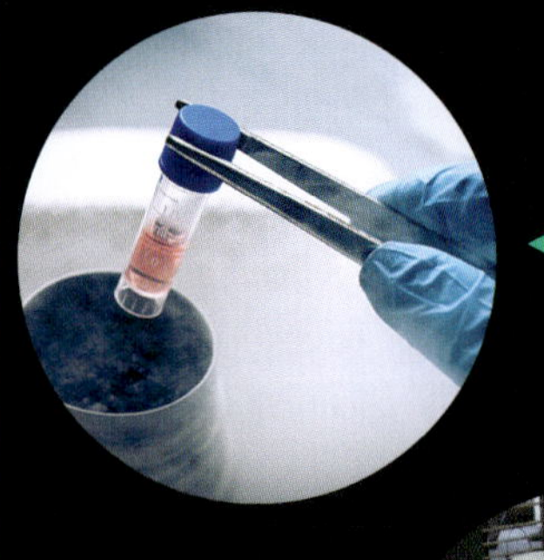

Stem cells

Cloning

Genetics and IVF technology

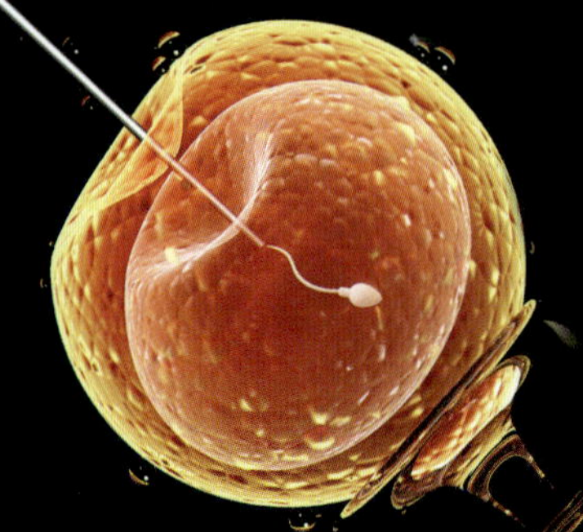

DNA replicates every time a new cell is formed.

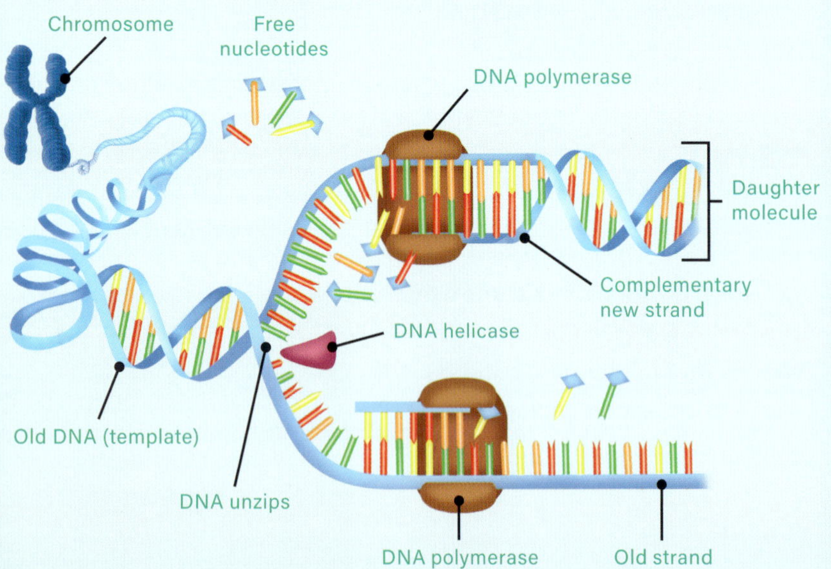

★ FINAL CHALLENGE ★

1. How many chromosomes do humans typically have?
2. Describe how genes are passed down from parents to their offspring.
3. Which DNA base pairs with:
   a adenine?
   b cytosine?
4. Explain what karyotypes are used for.

Level 1
50xp
LEVEL UP!

5. Explain what biotechnology is and give three examples.
6. Describe three types of mutation that can occur in DNA.
7. Explain the difference between a gene and a genome.
8. Explain what meiosis and fertilisation are in your own words.

Level 2
100xp
LEVEL UP!

9. Draw a diagram that shows the relationship between DNA, genes and chromosomes.
10. Describe some advantages and disadvantages of using stem cells in science and medicine.
11. Label each of the following as either homozygote recessive, homozygote dominant or heterozygote: AA, Bb, cc, Dd, Ee, FF.

12. Draw Punnett squares for the following parents.
    a Bb x bb
    b AA x aa
    c Tt x TT
13. Outline some ethical considerations of in vitro fertilisation.

14. Describe, using diagrams, how DNA is replicated. Suggest how mutations or errors can occur in this process.
15. Create a pedigree chart that shows two grandparents who have six children – four male and two female. Both of the men have two children of their own; one of them has twins.
16. Explain with the use of an annotated diagram how genes are passed from one generation to the next.

EVOLUTION

What makes us who we are?

Evidence suggests Earth is more than 4.5 billion years old. After 1 billion years, signs of life appeared on Earth. Tiny cells in the ocean gave rise to simple marine plant life, followed by land plants, and then small animals. Billions of years later, complex organisms live on Earth. These complex organisms can reproduce sexually to form genetically complex and intricate life, such as human beings. Evolution is the change in characteristics of living things over successive generations, often as a result of natural selection.

1 FRAYER MODEL

Copy and complete the below chart in your workbook.

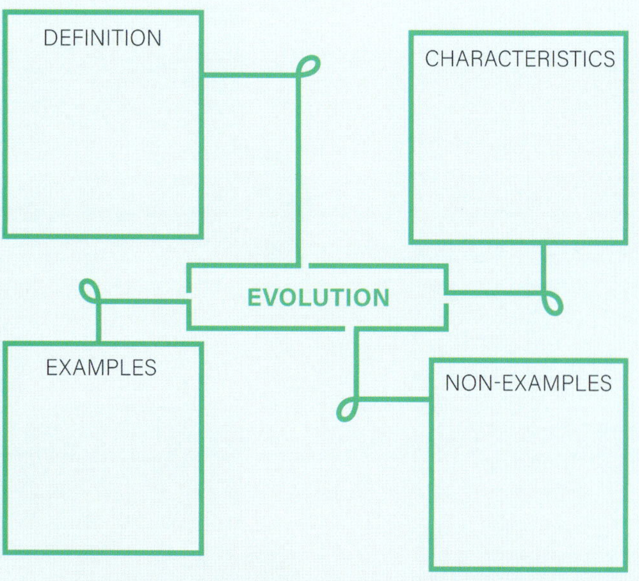

Complete two additional charts for the key terms *Biodiversity* and *Natural selection*.

2 LEARNING LINKS

Brainstorm everything you already know about evolution.

3 SEE-KNOW-WONDER

List three things you can SEE, three things you KNOW and three things you WONDER about this image.

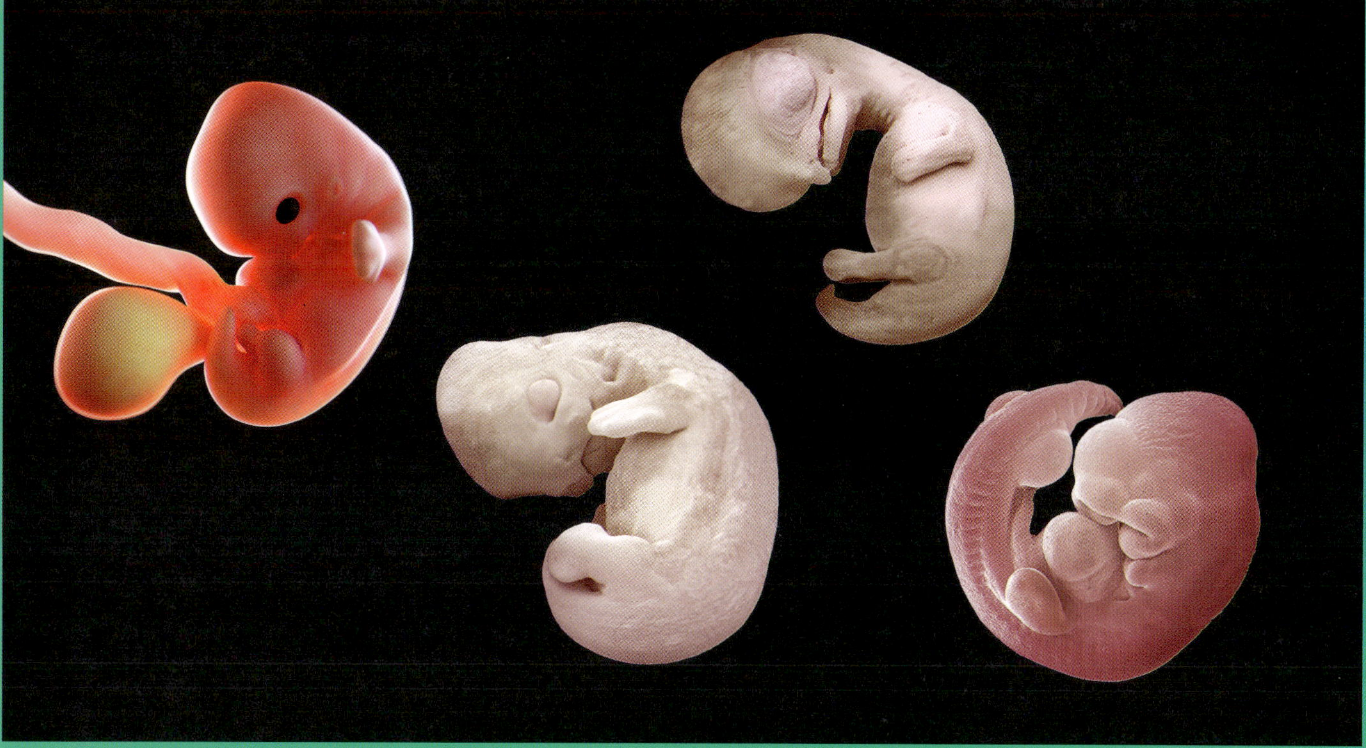

4 CRITICAL + CREATIVE THINKING

PREDICTION: What will a human look like in a million years?

INTERPRETATIONS: A fossil is found of a dinosaur with a human baby on its back. Write some explanations for this.

MISMATCHES: How could you solve the problem of decreasing biodiversity by using a fishing net, a kettle and a box of matches?

5 THE FLUFFIEST MONSTERS!

Scientists studying ancient fossils have made the rather surprising discovery that many dinosaurs (including *Tyrannosaurus rex*) probably had feathers. *T. rex* didn't use the feathers to fly; instead, the feathers probably helped them keep warm, socialise or even attract mates. Some dinosaurs' feathers are thought to have been all the colours of the rainbow.

2.1 BIODIVERSITY

LEARNING INTENTION

At the end of this lesson I will be able to describe biodiversity as a function of evolution.

KEY TERMS

biodiversity
the variety of living species on Earth

evolution
the gradual change in heritable characteristics over time

heritable
a characteristic that can be passed down from parent to offspring

LITERACY LINK

WRITING

Write a persuasive essay about a species that is classed as *endangered* or *threatened* and what should be done to ensure its survival.

NUMERACY LINK

GRAPHING

Graph the remaining numbers of the following species in a bar chart: Brazilian Spix's Macaw (100), Salt Creek Tiger Beetle (400), Amur Leopard (102), Javan Rhino (60).

Scientists have estimated that there are more than 8 million different species of animals, plants, bacteria and other living things on Earth. 8 million! And every single one has evolved over time from a common ancient ancestor. How did all these different species come to be? Why is the diversity of living things so important, and how can we ensure human activity does not lead to the extinction of so many of these species?

1 Evolution is the gradual change in heritable characteristics over time

Evolution is the way in which **heritable** characteristics are passed down over successive generations, causing change in living things over time. The theory of evolution was first proposed by Charles Darwin and Alfred Russel Wallace in the mid-1800s. They proposed that living things with characteristics best suited to their environment were more likely to survive. Think about it: if an organism can run faster, or is stronger or less impacted by disease, it is more likely to survive and therefore more likely to live to breed and pass these genes on to its offspring. Over a very long time, this results in the gradual evolution (change in characteristics) of the species. If a change occurs to an organism in its lifetime that is not passed down to its offspring, this is not considered to be part of evolution.

What is evolution?

Figure 2.1 Humans gradually evolved from apes over millions of years.

2 Biodiversity is a function of evolution

Biodiversity is the variety of living species on Earth. Considering that all life originated from a common ancestor, it is incredible how many different species inhabit Earth now. There are more than 900 000 different types of insects and more than 10 000 different types of birds, and new species are being discovered all the time. Over time, each of these species has developed unique characteristics. When an organism evolves so much that it can no longer successfully breed with its original species, it has become its own new species. Biodiversity is also proof of evolution; the amount of

Figure 2.2 There are more than 17 000 different species of butterflies on Earth.

variety that exists has come about because of evolution. If evolution didn't occur, all the different species on Earth would not exist because they would never have changed from their original life forms.

What is the relationship between biodiversity and evolution?

❸ Biodiversity is incredibly important to all life on Earth

Many people are extremely concerned about the rate of extinction on Earth. Most of this is due to human activities; for example, resource use including logging, fishing and diversion of natural waterways, as well as clearing land for agriculture (farming) and infrastructure (building).

Biodiversity is important for a number of reasons.

1 Humans use a huge variety of living things for food and other purposes.
2 Oxygen from plants, clean air and water is essential to all life.
3 Economically, many individuals and businesses rely on living things to buy or sell.
4 Many cultures, including Indigenous Australians, have deep and complex ties and spiritual beliefs in relation to the natural environment.
5 Alterations to biodiversity have flow-on effects for all other living things (remember food chains and food webs?).

What are some threats to biodiversity?

Figure 2.3 For tens of thousands of years, Aboriginal and Torres Strait Islander people lived harmoniously with Australia's natural environment.

CHECKPOINT 2.1

1 Explain evolution in your own words.
2 Which famous scientists have been credited with first proposing the theory of evolution?
3 Describe what biodiversity is, with examples.
4 A change that occurs to an individual in its lifetime that is not passed down to the next generation is not considered an example of evolution. Suggest why.
5 Give some reasons why biodiversity is important, including one that is not mentioned in this lesson.
6 What does 'biodiversity is a function of evolution' mean?
7 Give three examples of threats to biodiversity.

PERSONAL AND SOCIAL CAPABILITY

8 Working in a group of three, each group member will consider a different perspective – economic, environmental or cultural – on making use of the natural environment. Come up with some points from your chosen perspective, then respectfully listen to each other's opinions and negotiate a shared belief about the topic.

SUCCESS CRITERIA

- I can explain what evolution is in my own words.
- I can explain how biodiversity is related to evolution.

2.2 NATURAL SELECTION

LEARNING INTENTION

At the end of this lesson I will be able to explain natural selection and outline some of the processes involved.

Have you ever wondered why giraffes have such wonderfully long necks? Their long necks let them eat the highest leaves that other animals can't reach. Most scientists now think that giraffes got their long necks through a process called natural selection. A giraffe with a longer neck than other giraffes was much more likely than other giraffes to find food and survive. So, giraffes with long necks would be more likely to reproduce and pass their longer necks on to the next generation.

KEY TERMS

isolation
the different adaptations of members of a species that have been separated from the rest of the population

natural selection
the process whereby organisms better suited to their environment tend to survive and reproduce, passing on traits to future generations

speciation
the creation of new and unique species

variation
the natural differences between individuals in a population

1 Natural selection involves passing beneficial traits on to offspring

Natural selection occurs when beneficial traits are passed on from one generation to the next and is often referred to as 'survival of the fittest'. The beneficial traits are hereditary – they are passed on through the genes. However, factors other than 'fitness' affect natural selection – even luck.

Imagine the members of an insect species can be green or brown. Brown insects are better camouflaged on the bark of trees. Green insects are much more easily seen by predators, and so more green insects die. Over time, fewer green insects are left to reproduce, and eventually no green insects are left in the species – this is natural selection.

Natural selection was first proposed in the 1850s by celebrated scientist Charles Darwin, who put his ideas about natural selection and evolution into his book *On the Origin of Species*.

What is natural selection?

Figure 2.4 How many moths can you see here? The paler moth is camouflaged and so is much more likely than the black moth to survive and pass its genes on.

NUMERACY LINK

GRAPHING

A student measures the number of red bugs and green bugs on a bush as follows:

Time (days)	Red bugs	Green bugs
0	22	22
15	15	26
30	9	31
45	3	35

Graph this data and make a conclusion in terms of natural selection.

2 Processes involved in natural selection

There are a number of processes involved in natural selection, including **variation** (or selection) and **isolation**. Variation refers to the natural differences that exist between individuals in a population. These little differences can result in advantages that make certain individuals better suited to their environment and thus more likely to survive and reproduce. Variation is also known as selection. Isolation is when members of a species become removed and isolated from the rest, and over time they develop their own unique characteristics as they adapt to their different environment and become their own species. This is also known as **speciation**.

What is the cause of speciation?

INVESTIGATION 2.2
Natural selection in practice

KEY SKILL
Representing data to identify patterns and trends

▶ *Go to page **140***

3 Galapagos finches helped Darwin develop his theory of natural selection

In 1831, Charles Darwin set sail from England aboard the *HMS Beagle* as the ship's naturalist. On his journey, Darwin found enormous fossils and many species of animals that forced him to consider new theories for the origins of life on Earth.

During Darwin's five-year expedition, he discovered a number of species of finches in the Galapagos Islands, off the coast of Ecuador. Although the finches looked very similar, their beaks were different lengths and shapes, which reflected the different food sources on each island. Darwin theorised that the finches had a common ancestor but, after they were separated geographically and began eating different food sources, the finches evolved different beaks through natural selection.

What did Darwin learn about evolution from the finches in the Galapagos islands?

Figure 2.5 The finches of the Galapagos islands demonstrated the evolution of different beaks in response to the food sources available to them.

Large ground finch
(seeds)

Cactus finch
(cactus fruit and flowers)

Vegetarian finch
(buds)

Woodpecker finch
(insects)

CHECKPOINT 2.2

1 Define natural selection.
2 Who was Charles Darwin and what did he contribute to modern science?
3 Outline some key processes involved in natural selection.
4 How could being geographically isolated cause speciation?
5 How long do you think it would take for species to differentiate and evolve as Darwin's finches did? Justify your answer.
6 Natural selection is when beneficial traits are passed on to the next generation. Can you think of a situation where this wouldn't happen?

CONNECTING IDEAS

7 Consider the diagrams of Darwin's finches in Figure 2.5. Create a summary of why the form of each beak suits the function (accessing each food source).

SUCCESS CRITERIA

- I can define natural selection.
- I can describe some of the key processes involved in natural selection.

2.3 FACTORS AFFECTING SURVIVAL

LEARNING INTENTION

At the end of this lesson I will be able to outline the roles of genes and environmental factors in the survival of organisms in a population.

KEY TERMS

adaptation
a change in the structure or function of an organism, evolved over generations, that makes it better suited to its environment

genetic variation
the differences in genes within individuals of a population

LITERACY LINK

WRITING

Write a letter to the editor of a newspaper. In your letter, outline how habitat destruction affects the survival of plants and animals in Australia and suggest a way to combat habitat destruction.

NUMERACY LINK

CALCULATION

A scientist wants to measure the fish population in a 50 m^2 pond. He chooses five different areas of 1 m^2 each. His results are: 6, 4, 9, 3, 11. Calculate the approximate number of fish in the entire pond.

Many factors affect the survival of organisms in a population, including an individual's genetics (and corresponding physical traits) and the environment it is in.

Figure 2.6 Angler fish have many genetic adaptations for survival. Many environmental factors affect them too.

The angler fish (Figure 2.6) has many physical traits that have evolved over time and that allow the fish to survive in some of the darkest environments on Earth. Food is scarce in these environments, so the fish lures its prey with a little light projected in front of its huge mouth. Female angler fish release powerful chemicals so that potential mates can find them in the pitch-black waters.

1 Survival depends on genetic diversity in populations

DNA is the blueprint for your biology. DNA is found in the nucleus of most cells. Segments of DNA are called genes and can determine different physical characteristics. For example, your height and bone structure are partly determined by your genes, which were passed down to you from your mother and father.

Natural selection acts to produce a population that is well suited to an environment. Over time it can reduce genetic variation, with organisms becoming more genetically alike. **Genetic variation** describes the variation in DNA sequences that is present in individuals of the same species. Genetic variation is critical to the continuation of life on Earth.

A population is a group of the same species living in a specific area and interbreeding. When populations lack genetic diversity, they are vulnerable to changes in their environment. They may be unable to adapt to changes because they don't contain the necessary genetic options.

For example, the Tasmanian devil lacks genetic diversity because of its low numbers, isolation and generations of inbreeding. When a highly contagious facial tumour spread through the population in the mid-1990s, no individuals had genetic resistance to the condition. The disease now affects more than 80% of Tasmanian devils.

Figure 2.7 The Tasmanian devil has low genetic diversity and so is susceptible to contagious facial tumours.

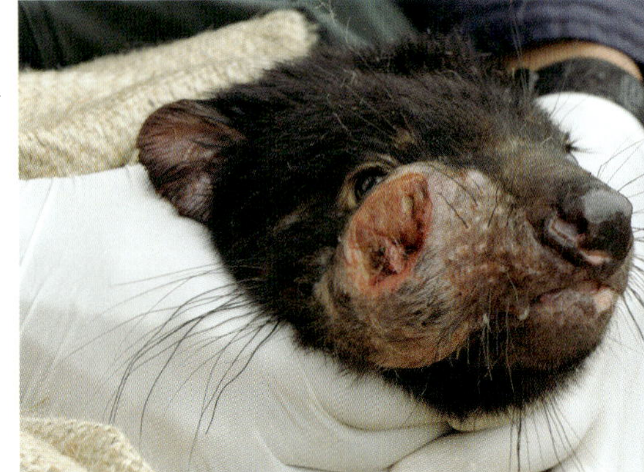

How can genetics influence survival?

❷ Environmental factors affect survival

Many environmental factors influence survival; for example, the amount of sunlight, temperature, wind, soil quality, pH and water. Most organisms can only survive in a fairly narrow temperature range. For example, polar bears thrive in freezing temperatures but would perish in tropical climates. Polar bears have several **adaptations** to their cold environment, including a thick coat and a layer of fat, as well as white fur that helps them blend in with their icy environment.

Other factors that affect survival are the availability of resources and mates, as well as the presence of predators and population density. Population density is a measure of the number of individuals living in a particular area and is very important to the survival of organisms. If the population density is too high, then individuals have to compete for food, water and reproductive mates. However, if the population density is too low, the population can lack genetic variation and it can be hard for individuals to find a mate.

How can the environment influence the survival of organisms?

❸ Human activity affects the survival of other species

Human civilisation has had a devastating effect on the natural environment and, thus, the survival of plants and animals. Habitat destruction has increased as humans have cleared land for farming, buildings and roads. Humans use fossil fuels, cut down trees for buildings or burning, build dams and change or destroy waterways.

Habitat destruction leaves plants and animals without the resources they need to live and reproduce. The World Wildlife Fund estimates that, since 1970, humans have caused the death of approximately 60% of all mammal, bird, reptile and fish species.

Conservation management can do much to assist species to avoid extinction. National parks provide protected places for plants and animals. Reducing deforestation, using natural resources sustainably and protecting habitats will safeguard some of the most vulnerable species.

How do human impacts on the environment affect the survival of organisms?

Figure 2.8 Habitat destruction has a devastating effect on the survival of plant and animal populations.

INVESTIGATION 2.3

Environmental and genetic factors affecting survival

KEY SKILL
Identifying the variables and formulating a hypothesis

▶ *Go to page **142***

CHECKPOINT 2.3

1 What is genetic variation?
2 Explain what population density is and how it can impact the survival of organisms in a population.
3 What role do genes play in the survival of an organism in a population?
4 List some environmental factors that affect the survival of organisms.
5 Explain why a lack of genetic variation can be very detrimental to the survival of a species.
6 Describe habitat destruction and how it can affect survival.
7 Suggest some ways to control and reduce habitat destruction.
8 Choose an animal and describe the genetic adaptations (physical traits) that improve its chances of survival.

EXTENSION

9 Research an endangered species. Write a short report summarising the threats to its survival.

SUCCESS CRITERIA

- I can explain the role of genes in the survival of organisms.
- I can explain the role of environmental factors in the survival of organisms.

2.4 EVIDENCE OF EVOLUTION

LEARNING INTENTION

At the end of this lesson I will be able to describe scientific evidence for present-day organisms having evolved from organisms in the past.

KEY TERMS

biogeography
the study of the past and present distribution of living organisms

comparative anatomy
the study of features of different species to look for evidence of a common ancestor

homologous structures
parts of organisms that show evidence of a common ancestor

vestigial structures
parts of organisms that have lost some or all of their original function

LITERACY LINK

VOCABULARY

Suggest how to break down the word *biogeography* in order to make meaning from the term.

NUMERACY LINK

MEASUREMENT

In adults, the femur (thigh bone) is said to be approximately ¼ of a person's height. Measure the length of your femur and measure your height. Determine your femur-to-height ratio. Record class results and calculate the average.

One of the closest relatives to humans is thought to be the bonobo, a great ape found in regions of the Congo in Africa. Bonobos live to about 40 years of age and communicate vocally and through facial expressions.

How do we know that humans are related to the bonobo? There is much scientific evidence that humans and bonobos have evolved from a common ancestor. This evidence includes fossils, comparative anatomy, genetic evidence and biogeographical evidence.

1 Comparative anatomy looks at common features in different species

Evolution is a gradual change in the physical characteristics of organisms over many generations. The first definite clues of evolution came by comparing the anatomy (especially the bones) of species to find similarities and differences. Scientists using **comparative anatomy** have found evidence to show that different species have evolved from the same ancestor. Parts of the body in different species that show evidence of common features are called **homologous structures**.

Vestigial structures are another type of evidence. These are parts of organisms that have lost their usefulness as a species evolved. They remain in the organism, sometimes even causing problems. Vestigial structures in humans include wisdom teeth, the appendix, the tailbone and nipples on men.

How are homologous structures evidence of evolution?

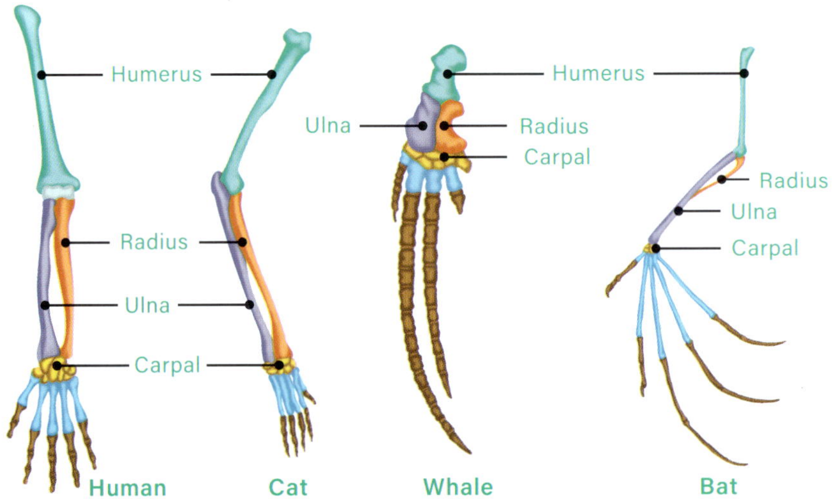

Figure 2.9 The human arm, cat leg, whale fin and bat wing are homologous structures that show evidence of a common ancestor.

❷ Genetic evidence shows how closely related different species are

DNA (deoxyribonucleic acid) is the molecule that codes for genetic instructions. DNA is a long molecule that contains all the biological information for humans and nearly every other living organism, and is found in nearly every cell of your body. The molecule forms a double helix – a twisted ladder of genetic information.

From 1990 to 2003, teams of scientists from all over the world worked together to completely map the DNA of humans in a huge endeavour called the Human Genome Project. Now we can compare the DNA of different species and identify how similar they are. DNA mapping shows that humans and bonobo great apes share more than 98% of their genes, providing evidence that we had a common ancestor.

How can DNA be used as evidence of evolution?

❸ Biogeography is used as evidence of evolution

Biogeography is the study of the past and present distribution of living organisms. By studying where plants and animals are or were found on islands and continents, scientists can start to piece together a complex and ancient puzzle.

Much of the evidence for biogeography comes from fossils. If fossils of one species are found in both South America and Africa, then this is compelling evidence that the two continents were once connected.

Australia has been an isolated island for many millions of years, which is why Australian marsupials and mammals are unique. However, even after being separated from South America for millions of years, Australian marsupials still have some common features with those in South America. This means that they probably evolved from a common ancestor millions of years ago.

How can biogeography be used as evidence of evolution?

Figure 2.10 DNA is contained in a structure like a twisted ladder called the double helix.

INVESTIGATION 2.4
Comparative anatomy

KEY SKILL
Explaining results using scientific knowledge

▶ *Go to page **144***

CHECKPOINT 2.4

1 Describe evolution.
2 What is biogeography and what can it tell us about the evolution of species?
3 Explain how the Human Genome Project led to breakthroughs in using genetics to show evidence of evolution.
4 Outline the evidence that suggests that Australia has been an isolated island for many millions of years.
5 Describe the difference between homologous and vestigial structures.
6 Which of biogeography, genetics or comparative anatomy do you think gives the most scientific evidence of evolution? Justify your response.

CONNECTING IDEAS

7 Embryology is the study of embryos. Embryos of many different kinds of animals look very similar. How might the study of embryology provide evidence of evolution?

SUCCESS CRITERIA

- I can give specific examples of scientific evidence that supports evolution.
- I can describe what is meant by comparative anatomy, genetic evidence and biogeography.

2.5 THE FOSSIL RECORD

LEARNING INTENTION

At the end of this lesson I will be able to relate the fossil record to the age of Earth and the time over which life has been evolving.

KEY TERMS

fossil
the geologically altered remains of an organism that was once alive

fossil record
a record of all fossils found on Earth

palaeontologist
a scientist who studies fossils

LITERACY LINK

WRITING

Write a job advertisement for a palaeontologist. Your advertisement should include a description of the work involved and a typical day in the life of a palaeontologist.

NUMERACY LINK

GRAPHING

Use the dates in Figure 2.13 to construct a pie chart showing the relative durations of each of the different eras: Caenozoic, Mesozoic and Palaeozoic.

When you hold a fossil, you are holding a piece of ancient history, a remnant from millions of years ago when Earth was young and nearly unrecognisable.

Many things can be turned into a fossil, but the conditions must be perfect, which is why fossils are so rare. Fossils contain clues to the history of life on Earth and are important in the scientific study of comparative anatomy, including our own complex heritage.

Figure 2.11 Fossils form when layers of sediment cover an organism. After millions of years, the organism can become an impression in solid rock.

1 Fossils form over millions of years under the right conditions

Fossils are the remains, impressions or traces of long dead organisms left in rock and kept safe by geological processes. Fossils can be preserved bones or other remnants, shells, footprints, nests and eggs. Fossils are rare because they only form under certain conditions – the temperature, amount of oxygen and even soil acidity must be just right.

First, the organism must be protected once it dies so that its bones (or seeds and bark) stay together and in relatively good condition. This happens much more easily under water because the organism sinks to the bottom and is covered by mud or silt. Over many years, layers of sediment cover the organism.

Over millions of years, the sediments harden and become rock, and the organism is then preserved as a fossil.

Fossils can also form when the bones completely decay to leave a hollow mould. Other minerals then seep in and harden, making a three-dimensional replica of the original bones.

How do most fossils form?

1. Remains of organism in water

2. Remains buried by layers of sediment

3. Remains are hardened and become rock

4. Remains become exposed on surface as fossil

Figure 2.12 Fossils form over millions of years, usually in sedimentary rock.

❷ The fossil record provides evidence of evolution

Palaeontologists study the **fossil record**, which provides evidence of evolution. It shows how organisms have changed through geologic time from the simplest single-cell organisms to human beings.

Fossils have been found and dated from all around the world but it is thought many species are missing from the record, or have not yet been discovered. It is hard to get an exact picture of all life forms from the fossil record because of the gaps. Also, it is hard to know whether a particular fossil is a good representation of a species.

What is the fossil record?

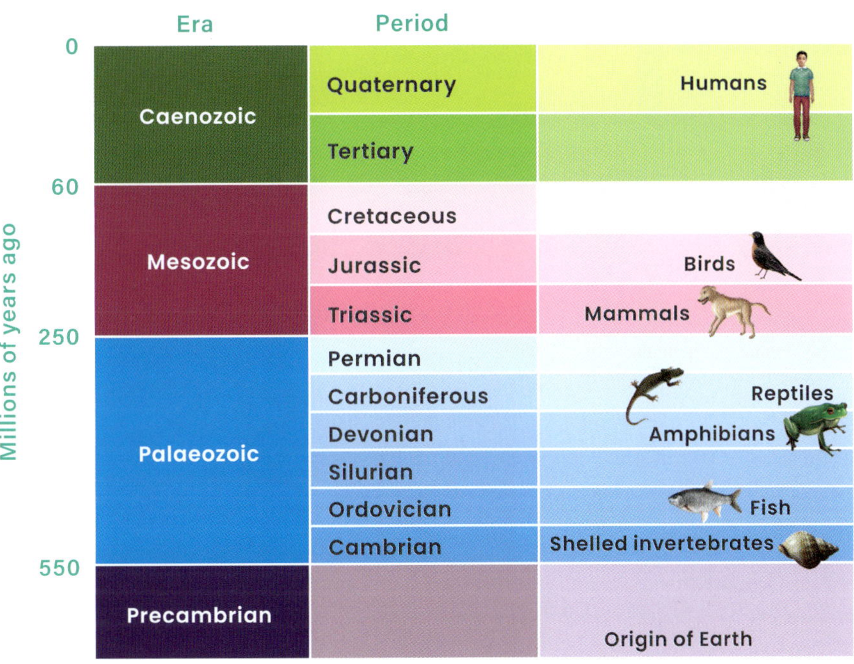

Figure 2.13 This sequence of animal fossils begins with 500-million-year-old fossils of simple invertebrates and ends with human beings.

❸ Fossils tell us about the age of Earth and evolution

Scientists date fossils in two main ways – relative and absolute dating. Relative dating gives an approximate age of a fossil by comparing it to other fossils or by studying the layers and ages of the surrounding stone. Absolute dating is commonly done by studying the radioactive minerals in the rock. Scientists can calculate the age of the rock by investigating how much the radioactive elements in them have decayed. Older fossils are of simpler organisms than younger fossils. This evidence supports the theory of evolution because it suggests that much more complex organisms evolved from simpler ones.

How can fossils be used as evidence of evolution?

INVESTIGATION 2.5
Dating fossils

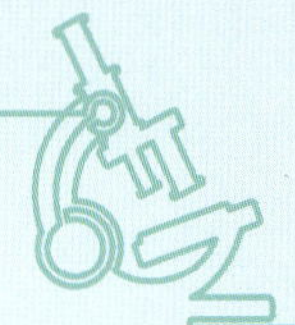

KEY SKILL
Using modelling and simulations

▶ *Go to page* ***145***

CHECKPOINT 2.5

1 What is the fossil record?
2 Give two reasons why the fossil record is incomplete.
3 Explain the difference between relative and absolute dating.
4 The oldest fossils discovered are of the simplest organisms. Explain why this is the case.
5 Create a flow chart showing the process of fossilisation.
6 Explain how fossils can be used as evidence of evolution.
7 Outline the two main ways to establish the age of a fossil.
8 Fossils occur mostly in sedimentary rock. Suggest why this might be the case.
9 Give some reasons why the fossil record may be viewed as an 'incomplete puzzle'.

RESEARCH

10 Research the biggest fossil discoveries made by scientists. Choose one that you think is the most significant scientifically and explain why.

SUCCESS CRITERIA

- I can describe how a fossil is formed.
- I can explain why the fossil record can be used as evidence of evolution.

VISUAL SUMMARY

Evidence of evolution

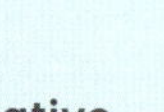

- Comparative anatomy
- Genetics
- Biogeography

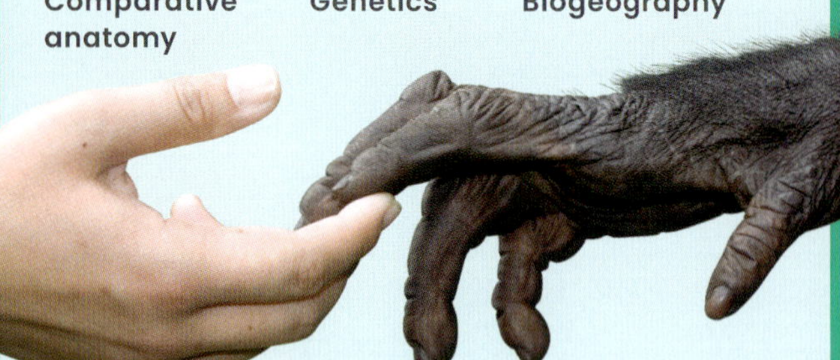

Survival relies on:

▲ genetic diversity

▲ human activity

▼ environment

▼ Fossils form over millions of years.

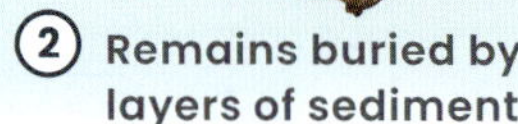

1. Remains of organism in water

2. Remains buried by layers of sediment

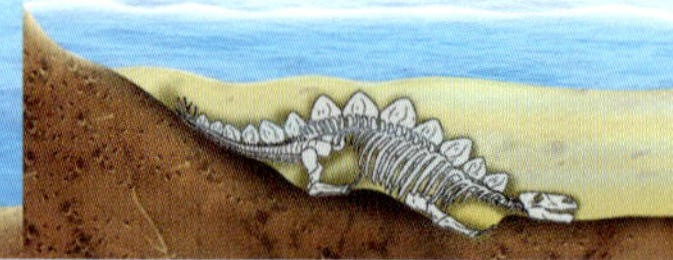

3. Remains are hardened and become rock
4. Remains become exposed on surface as fossil

▼ The fossil record provides evidence of evolution.

Millions of years ago	Era	Period	
0	Caenozoic	Quaternary	Humans
		Tertiary	
60	Mesozoic	Cretaceous	
		Jurassic	Birds
		Triassic	Mammals
250	Palaeozoic	Permian	
		Carboniferous	Reptiles
		Devonian	Amphibians
		Silurian	
		Ordovician	Fish
		Cambrian	Shelled invertebrates
550	Precambrian		Origin of Earth

Natural selection

'Survival of the fittest' → Beneficial traits passed down → Populations become better suited to their environment → 'Survival of the fittest'

The Galapagos finches demonstrate evolution by natural selection, developing different beaks according to the food sources available.

Large ground finch (seeds)

Cactus finch (cactus fruit and flowers)

Vegetarian finch (buds)

Woodpecker finch (insects)

★ FINAL CHALLENGE ★

1. Define evolution.
2. Describe how fossils are formed.
3. Explain how environmental factors affect the survival of individuals and populations.
4. Describe biodiversity in your own words.

Level 1
50xp
LEVEL UP!

5. Explain how competition is an environmental factor that affects survival.
6. Describe natural selection and give an example.
7. Explain how Darwin's finches demonstrate principles of natural selection.
8. Explain the relationship between evolution and biodiversity.

9. Fossils are quite rare. Suggest why.
10. Explain how fossils and comparative anatomy are linked to our understanding of evolution.
11. Explain how natural selection has resulted in bacteria that are resistant to antibiotics.
12. Suggest ways to stop or slow habitat destruction.

13. Analyse the usefulness of the fossil record in providing evidence of evolution.
14. Which do you think is more important to the survival of organisms – genes or environmental factors? Give evidence to support your opinion.
15. Write three journal entries from Charles Darwin's perspective during his five-year expedition.

16. Define the following terms and outline the evidence they provide of evolution.
    - a comparative anatomy
    - b genetics
    - c biogeography
    - d fossils
17. Explain, using examples, the processes of variation, isolation and selection.

THE PERIODIC TABLE

How does the chemical world shape our lives?

The periodic table is a way of sorting elements into groups based on their physical and chemical properties.

The left side of the periodic table contains some very reactive elements, such as hydrogen, cesium and potassium. Pure cesium and potassium have to be stored in oil so they don't come in contact with any other element. Otherwise, they quickly catch fire and explode.

As you move towards the centre of the periodic table, the elements tend to become more stable. However, as you move past the centre and to the right, they become increasingly unstable again. Fluorine, for example, is the most reactive non-metal element in the whole table. The last group, the noble gases, is the least reactive group in the whole table.

1 FRAYER MODEL

Copy and complete the below chart in your workbook.

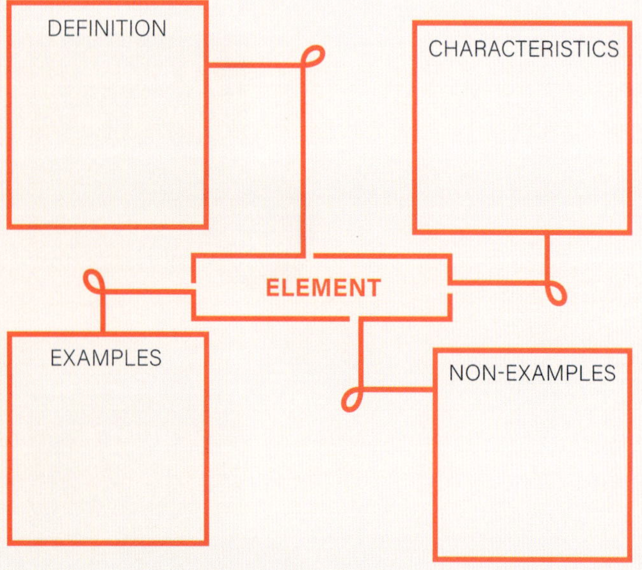

Complete two additional charts for the key terms *Metal* and *Electron*.

2 LEARNING LINKS

Brainstorm everything you already know about the periodic table.

3 SEE-KNOW-WONDER

List three things you can SEE, three things you KNOW and three things you WONDER about this image.

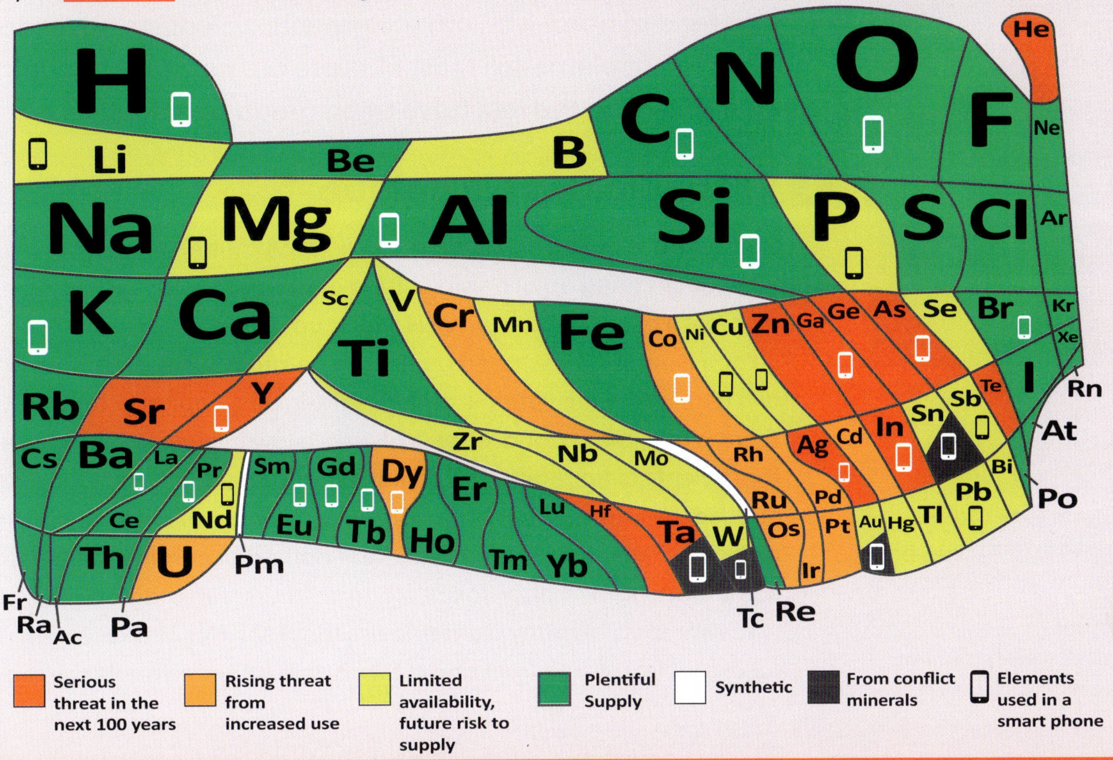

Serious threat in the next 100 years
Rising threat from increased use
Limited availability, future risk to supply
Plentiful Supply
Synthetic
From conflict minerals
Elements used in a smart phone

CRITICAL + CREATIVE THINKING

THE REVERSE: List five things that are NOT made of elements.

DIFFERENT USES: Think of as many different uses for helium as possible.

THE COMMONALITIES: List as many commonalities between carbon and an iceberg as you can think of.

THE MYSTERY OF YTTERBY!

Strangely, one of the most important places in the history of chemistry is a small village called Ytterby on a tiny Swedish island. The village had a quarry that was used for mining a mineral called feldspar. In this single quarry, seven previously undiscovered rare earth elements were found. Four of them were named after Ytterby itself – ytterbium, terbium, erbium and yttrium – and there was also holmium, thulium and gadolinium. Almost 6% of all of the elements ever discovered were found on an island you could walk around in a few hours.

3.1 ELEMENTS AND THEIR ATOMS

LEARNING INTENTION

At the end of this lesson I will be able to describe elements by comparing information about the number of protons, neutrons and electrons.

KEY TERMS

atom
the smallest particle of matter, made up of electrons orbiting a nucleus of protons and neutrons

chemical symbol
one or two letters that represent an element

element
a pure substance made up of one type of atom

ion
an atom with unequal numbers of protons and electrons; a charged atom

isotope
an atom of a particular element with a different number of neutrons

LITERACY LINK

WRITING

Construct the longest possible word using only element symbols, without repeating any.

NUMERACY LINK

GRAPHING

Draw a line graph for the first 20 elements, with their atomic number on the x-axis and atomic mass on the y-axis. Describe the trend of the graph.

All matter is made of different elements. These elements are made up of atoms, which we cannot see. Some elements are stable and exist on their own. Other elements are very unstable and tend to bond with other elements. Scientists have identified 118 different elements. Most of these are naturally occurring elements, and the rest have been made by scientists.

1 Every element is unique

Elements are the basic building blocks of all matter. An **element** is a pure substance made up of only one type of **atom** – the smallest unit of an element. For example, oxygen is made up of only oxygen atoms.

Each element is assigned a unique **chemical symbol** using one or two letters. The first letter is always written as a capital letter and, when there is one, the second letter is always a lower-case letter. This ensures that scientists from different countries, who speak different languages, still use the same symbols in their work. The names of elements have different origins. Some elements were named according to their properties, some were named after scientists, and others were named after places.

The symbols of many elements are the first letter of their English names; for example, O for oxygen and C for carbon. Other symbols are made up of two letters from their English name; for example, Cl for chlorine and Mg for magnesium. You will find these symbols easy to identify.

Other symbols are not as easy to identify because they come from the Latin, Greek or German names of the elements. These elements were discovered centuries ago. For example, the symbol for:

- gold (Au) comes from the Latin name *aurum*
- iron (Fe) comes from the Latin name *ferrum*
- sodium (Na) comes from the Latin name *natrium*
- mercury (Hg) comes from the Greek word *hydrargyros*
- tungsten (W) comes from the German word *wolfram*.

How are elements assigned a chemical symbol?

Figure 3.1 The element curium is named after physicist and chemist Marie Curie.

❷ Different elements have different numbers of protons

Each element has a unique number of protons. For example, every atom with 6 protons is carbon, while every atom with 7 protons is nitrogen.

However, the same element can have different numbers of neutrons and electrons. Carbon-12 is the most common form of carbon, and it contains 6 protons and 6 neutrons. Carbon-14 is rarer and contains 6 protons and 8 neutrons. Every single carbon atom will always have 6 protons – if it didn't, it wouldn't be carbon – but the number of neutrons within those atoms can change. Atoms of a particular element with different numbers of neutrons are called **isotopes**.

Similarly, atoms of the same element can have different numbers of electrons. A neutral atom has the same number of protons and electrons, whereas a charged atom (called an **ion**) has different numbers of protons and electrons. An oxygen atom has 8 protons, so a neutral oxygen atom has 8 electrons. However, an oxygen ion has 8 protons and 6 electrons.

Table 3.1 shows the first 10 elements, as well as the number of protons, the average number of neutrons and the number of electrons in a neutral atom.

What are elements?

Table 3.1 Composition of the most common forms of the first 10 elements

Element name	Symbol	Protons	Neutrons	Electrons
Hydrogen	H	1	0	1
Helium	He	2	2	2
Lithium	Li	3	4	3
Beryllium	Be	4	5	4
Boron	B	5	6	5
Carbon	C	6	6	6
Nitrogen	N	7	7	7
Oxygen	O	8	8	8
Fluorine	F	9	10	9
Neon	Ne	10	10	10

❸ The periodic table lists all of the elements

All of the 118 elements that science has discovered are shown on the periodic table.

The table is more than just a 1–118 list, though. Its structure groups together elements with similar properties. It also shows us how many protons each element contains.

The full periodic table is shown on the next two pages.

How many elements have been identified to date?

CHECKPOINT 3.1

1. Describe the difference between an ion and an isotope in your own words.
2. Use the periodic table on the next pages to find the chemical symbols for:
   a. sodium
   b. silver
   c. sulphur.
3. Explain why elements are given a chemical symbol and describe how these symbols might be chosen.
4. The number of protons, not the number of neutrons or electrons, is what makes an element unique. Explain why.

RESEARCH

5. Carry out research to identify the most unstable element on the periodic table. Describe the element and include how many protons, neutrons and electrons it has.

SUCCESS CRITERIA

- I can give an example of the number of protons, electrons and neutrons in at least one element.
- I can explain what an element is.

3.1 continued ...

... 3.1 continued

ELEMENTS AND THEIR ATOMS

Figure 3.2 The modern periodic table lists the elements that have been discovered to date.

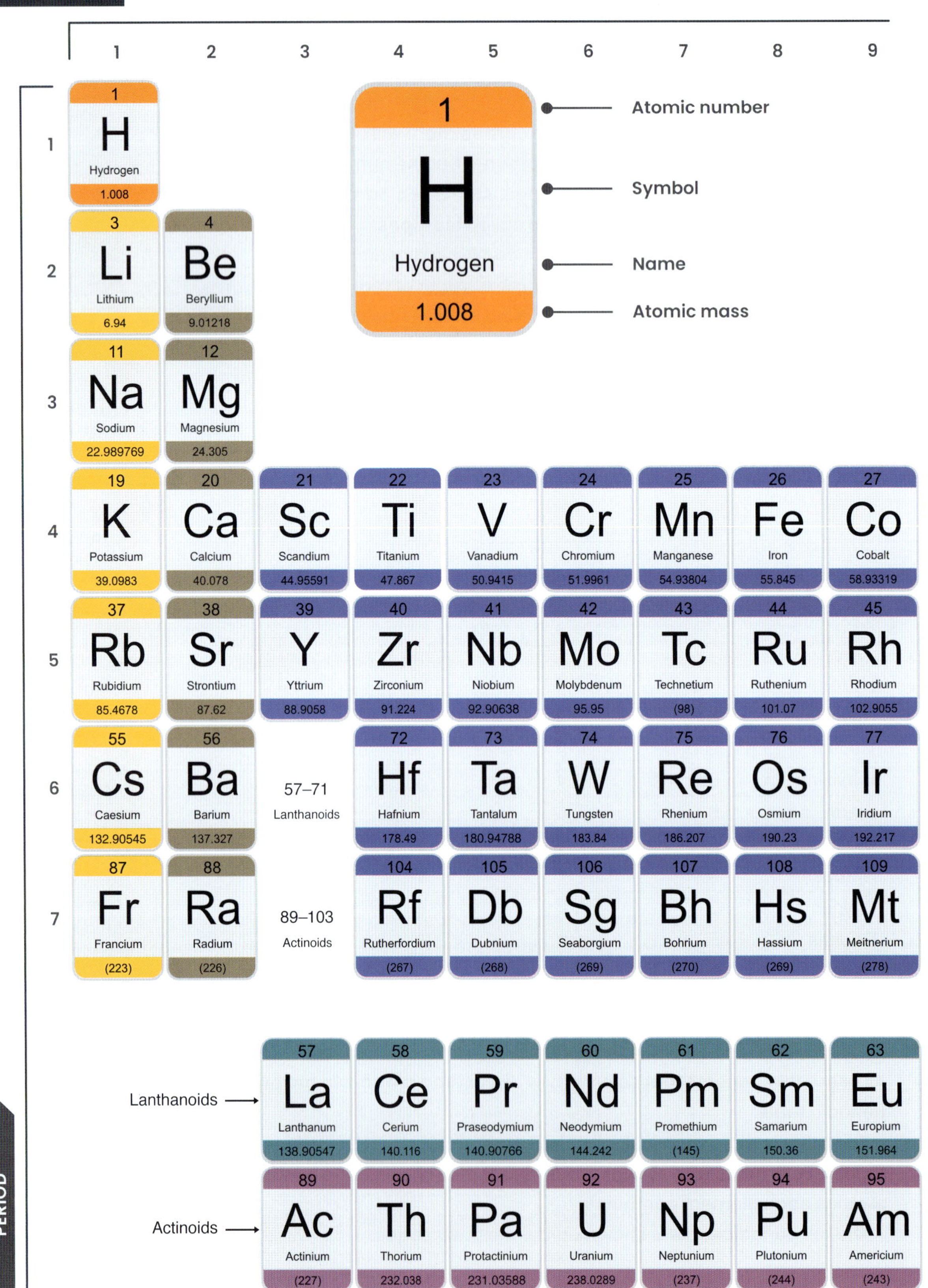

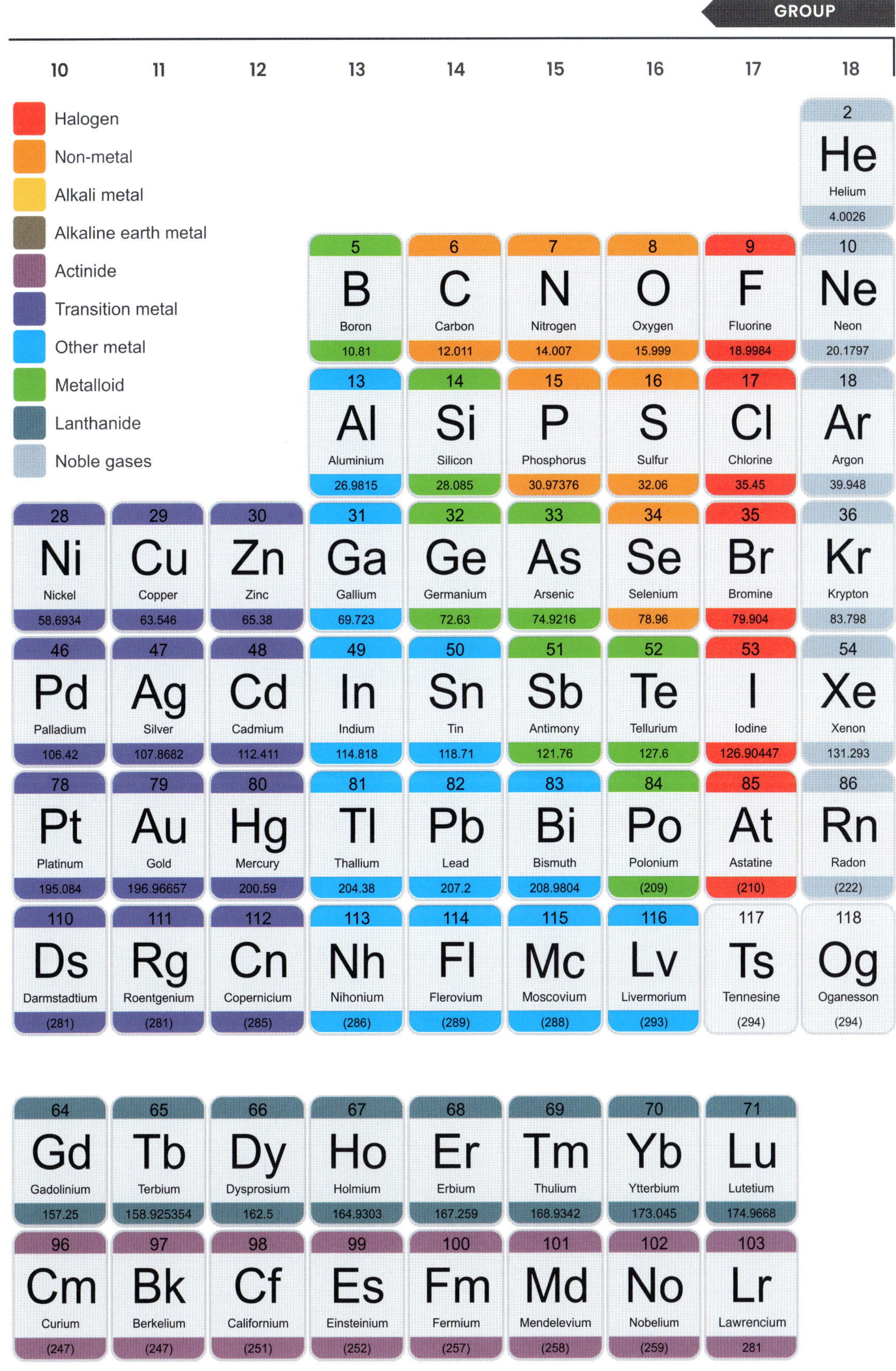
GROUP
10
11
12
13
14
15
16
17
18
Halogen
Non-metal
Alkali metal
Alkaline earth metal
Actinide
Transition metal
Other metal
Metalloid
Lanthanide
Noble gases
2 He Helium 4.0026
5 B Boron 10.81
6 C Carbon 12.011
7 N Nitrogen 14.007
8 O Oxygen 15.999
9 F Fluorine 18.9984
10 Ne Neon 20.1797
13 Al Aluminium 26.9815
14 Si Silicon 28.085
15 P Phosphorus 30.97376
16 S Sulfur 32.06
17 Cl Chlorine 35.45
18 Ar Argon 39.948
28 Ni Nickel 58.6934
29 Cu Copper 63.546
30 Zn Zinc 65.38
31 Ga Gallium 69.723
32 Ge Germanium 72.63
33 As Arsenic 74.9216
34 Se Selenium 78.96
35 Br Bromine 79.904
36 Kr Krypton 83.798
46 Pd Palladium 106.42
47 Ag Silver 107.8682
48 Cd Cadmium 112.411
49 In Indium 114.818
50 Sn Tin 118.71
51 Sb Antimony 121.76
52 Te Tellurium 127.6
53 I Iodine 126.90447
54 Xe Xenon 131.293
78 Pt Platinum 195.084
79 Au Gold 196.96657
80 Hg Mercury 200.59
81 Tl Thallium 204.38
82 Pb Lead 207.2
83 Bi Bismuth 208.9804
84 Po Polonium (209)
85 At Astatine (210)
86 Rn Radon (222)
110 Ds Darmstadtium (281)
111 Rg Roentgenium (281)
112 Cn Copernicium (285)
113 Nh Nihonium (286)
114 Fl Flerovium (289)
115 Mc Moscovium (288)
116 Lv Livermorium (293)
117 Ts Tennesine (294)
118 Og Oganesson (294)
64 Gd Gadolinium 157.25
65 Tb Terbium 158.925354
66 Dy Dysprosium 162.5
67 Ho Holmium 164.9303
68 Er Erbium 167.259
69 Tm Thulium 168.9342
70 Yb Ytterbium 173.045
71 Lu Lutetium 174.9668
96 Cm Curium (247)
97 Bk Berkelium (247)
98 Cf Californium (251)
99 Es Einsteinium (252)
100 Fm Fermium (257)
101 Md Mendelevium (258)
102 No Nobelium (259)
103 Lr Lawrencium 281

3.2 ELECTRON CONFIGURATION

LEARNING INTENTION

At the end of this lesson I will be able to describe the structure of atoms in terms of electron shells.

KEY TERMS

Bohr diagram
a diagram showing the electron configuration of an atom

electron configuration
the organisation of electrons into shells around the nucleus

electron shells
orbits around the nucleus that hold electrons of similar energy levels

valence electrons
electrons orbiting in the outer shell of an atom

valence shell
the outermost shell of an atom

LITERACY LINK

LISTENING

Think of an element and describe it to a partner in terms of its electron shells, while your partner tries to draw it from your description.

NUMERACY LINK

UNITS

The radius of orbit of an electron around a hydrogen atom is 53 picometres. Convert this into nanometres, then into metres. (1 picometre = 0.001 nanometres = 1 × 10^{-12} metres)

All atoms are made up of protons, neutrons and electrons. The protons and neutrons are always located together, in the nucleus (the very middle) of the atom, and so it is easy to determine where they are. The electrons are all located outside the nucleus, but different electrons will exist in different places within the atom. So how are we to tell where electrons are located? The answer is a concept called **electron configuration ,** which is illustrated by **Bohr diagrams**.

1 Electrons exist in shells

You have probably seen a stylised image of an atom that shows particles orbiting the centre of the atom, such as Figure 3.3. These particles orbiting the centre are electrons, and you can visualise them orbiting the nucleus in much the same way that planets orbit the Sun. The distance between the electrons and the nucleus comes down to energy – the more energy the electrons have, the farther they are from the nucleus. Some electrons have similar energy levels to each other, and so orbit at roughly the same distance. We call these orbits **electron shells**.

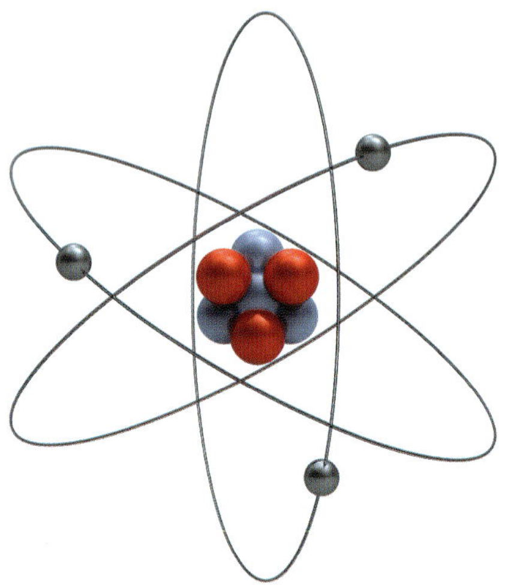

Figure 3.3 Electrons orbit the nucleus.

The farther from the nucleus a shell is, the more electrons it can hold. The maximum number of electrons in any shell is $2n^2$, where n is the shell number. This is shown in Table 3.2.

Table 3.2 Maximum number of electrons by shell number

Shell number	1	2	3	4
Maximum number of electrons	2	8	18	32

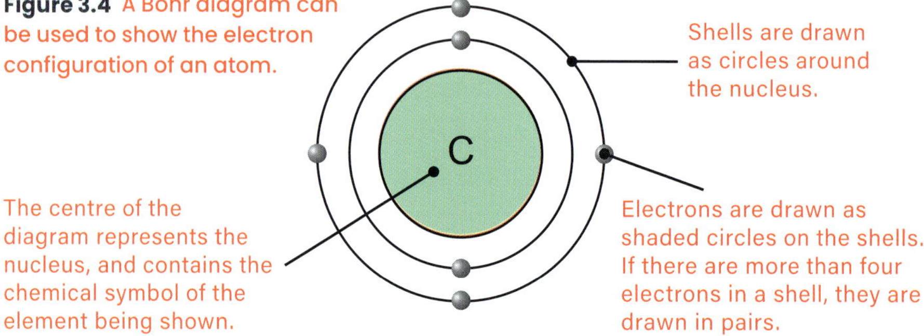

Figure 3.4 A Bohr diagram can be used to show the electron configuration of an atom.

What is an electron shell?

❷ Rules for drawing Bohr diagrams

1 The electrons fill up the lowest energy shells first (these are usually closest to the nucleus, so shell 1 fills up before shell 2, and so on).

2 The outer shell (called the **valence shell**) can never have more than 8 electrons in it.

When we consider the first 20 elements, the electron shells follow a 2, 8, 8, 2 pattern. That is, we fill the shells in order, except we move to the fourth shell once the third shell has 8 electrons in it. Elements larger than calcium don't follow this rule, so we will focus on smaller atoms. Two examples are below.

- Lithium has three electrons. The first shell can only hold two, and so the third electron will be in the second shell. This configuration is written as 2, 1.
- Potassium has 19 electrons. The first shell can hold two and the second shell can hold eight, so there will be nine left over after filling those first two shells. Remember, we put eight in the third shell before starting on the fourth, so there will be one electron in the fourth shell. This gives potassium an electron configuration of 2, 8, 8, 1. Adding up the number of electrons in each shell should give you the total number of electrons in that atom. For potassium, this is 2 + 8 + 8 + 1 = 19.

What is the valence shell?

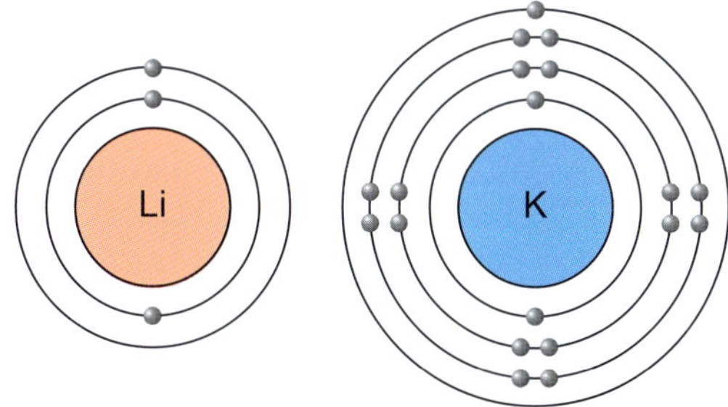

Figure 3.6 These Bohr diagrams represent the electron configurations of lithium and potassium.

❸ Chemical properties depend on valence electrons

The chemical properties of an element are directly related to the number of **valence electrons** that it has. For example, sodium and rubidium react very similarly to each other when combined with other elements – they both combust in contact with water and react very strongly with acids. Rubidium is a much larger atom than sodium, but their reactions are similar because they both have one valence electron. Likewise, carbon and silicon react very similarly, because they both have four valence electrons. The number of valence electrons is the most important factor in determining the chemical properties of an element.

What determines the chemical properties of an element?

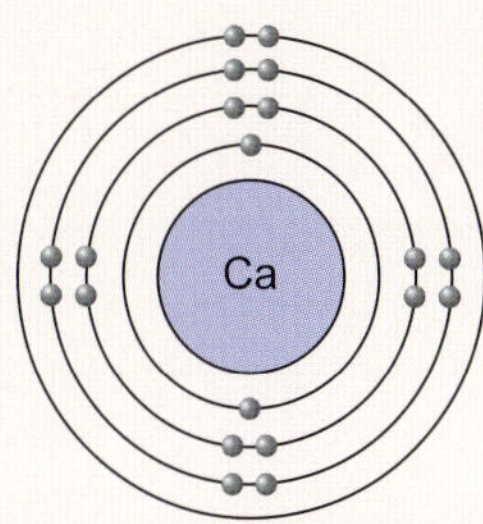

Figure 3.5 Calcium is the largest atom to follow the 2, 8, 8, 2 rule.

CHECKPOINT 3.2

1 Draw a Bohr diagram and write the electron configuration of:
   a sodium
   b aluminium
   c phosphorus.

2 Would electrons with more energy be closer or farther away from the atom?

3 Name three elements that would share similar chemical properties to oxygen. Explain why.

4 What would be the maximum number of electrons that the fifth shell could hold? How do you know?

5 A certain atom has an electron configuration of 2, 8, 7. What element is this atom?

6 You can add energy to electrons by heating them up. What do you think happens to an electron's location when it is heated up?

CRITICAL AND CREATIVE THINKING

7 Create a model of a calcium atom that you could use to teach someone else about electron shells. You can make your model digital or physical, using any resources you like.

SUCCESS CRITERIA

- I can determine the electron configuration of any of the first 20 elements.
- I can describe how energy is related to electron shells.

3.3 ORGANISING ELEMENTS

LEARNING INTENTION

At the end of this lesson I will be able to explain how the electronic structure of an atom determines its position in the periodic table and its properties.

KEY TERMS

atomic number
the number of protons in an atom

group
a column in the periodic table

mass number
the sum of the protons and neutrons in an atom

metalloid
an element with properties similar to both metals and non-metals

period
a row in the periodic table

LITERACY LINK

VOCABULARY

Choose five uncommon words from this spread. Using your own words, write definitions that a younger student could understand.

NUMERACY LINK

CALCULATION

The atomic number of potassium is 19 and its mass number is 39. Determine the number of protons, electrons and neutrons in a potassium atom.
Do the same for bromine, which has an atomic number of 35 and a mass number of 80.

A wide variety of substances exist in nature. These substances are made up of elements, which have different atoms. The periodic table is a way of classifying these elements. In the periodic table, elements are organised by their atomic number and grouped according to their properties. When the periodic table was being developed, scientists left gaps for elements that were still unknown at the time. They could predict the properties of the undiscovered elements by studying the properties of the surrounding elements.

1 Atomic number is the number of protons

The **atomic number** is the number of protons that each atom of the element has in its nucleus. The elements on the periodic table are arranged by increasing atomic number. In the periodic table, atomic number is usually written above the element's symbol.

For example, every atom of silicon (Si) has 14 protons, so the atomic number of silicon is 14. Similarly, every atom of oganesson (Og) has 118 protons so the atomic number of oganesson is 118.

How are elements organised in the periodic table?

2 Mass number is the number of protons and neutrons

The number of protons and neutrons in the nucleus is known as the **mass number**.

Mass number = number of protons + number of neutrons

The periodic table doesn't list the mass number of elements. Instead, it shows the atomic mass, which is the the average number of protons and neutrons for all natural isotopes of an element. This is always a decimal number.

How is mass number calculated?

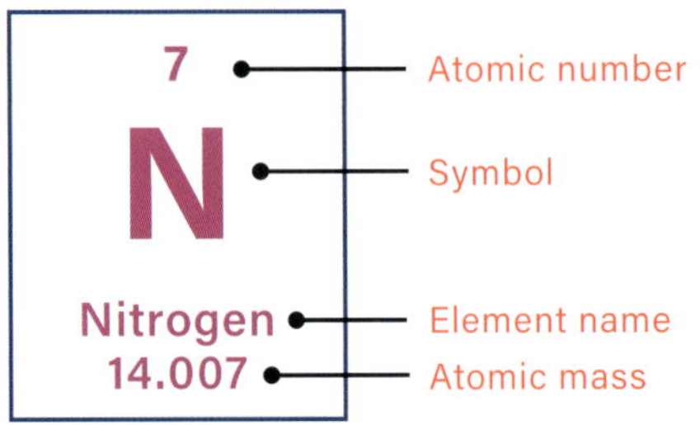

Figure 3.7 The periodic table includes information for every known element, including the element's name and symbol, atomic mass and atomic number.

❸ Elements in the same group share some properties

In the periodic table, elements are arranged into 18 **groups**, the vertical columns based on their physical and chemical properties. Groups indicate the number of valence (outermost) shell electrons – the number of the group is related to the number of valence electrons and chemical properties of the elements within it.

Group 1 metals are called the alkali metals and have one valence electron. They are very reactive and are soft, shiny, malleable and conduct heat and electricity. Group 2 metals are the alkaline earth metals and have two electrons in their valence shell. They have similar metallic properties to group 1 metals, except they are less reactive.

The middle of the periodic table contains the transition metals, which are all solids except mercury, which is liquid. To the right of the transition metals are the semimetals (or **metalloids**), such as boron, silicon, germanium, arsenic, antimony and tellurium. Their properties are a blend of metal and non-metal properties.

The non-metals are found on the right-hand side of the periodic table. Group 17 is known as the halogens, the most reactive non-metal group. Halogens have seven electrons in their valence shell and are all gases except bromine, which is liquid at room temperature, and iodine and astatine, which are solids.

Group 18 is known as the noble gases. They are also known as the inert gases because they are unreactive.

The horizontal rows are called **periods**, and they are numbered 1 to 7. Elements in the same period have the same number of electron shells, or orbits in which electrons can be found. Potassium (K) and bromine (Br) have very different properties, but they both have four electron shells and are found in period 4.

Two parts of the periodic table are broken out into separate rows. These are the lanthanoids (57–71) and the actinoids (89–103). These elements are rare and have very similar properties, so they're broken out into their own rows so that they don't disrupt the arrangement of the table.

What does an element's group and period tell us?

Figure 3.8 Noble gases are often used in neon lights.

CHECKPOINT 3.3

1 Explain why the atomic number of an element is a whole number, but the atomic mass is a decimal number.

2 Give two similarities and two differences between group 1 and group 2 elements.

3 Describe the elements found in the middle of the periodic table.

4 a Where are non-metals located in the periodic table?
  b How does the number of non-metals compare to the number of metals?

5 What is the difference between the mass number and atomic mass?

6 Explain what the groups and periods are in the periodic table.

7 Use the periodic table to find the atomic number of:
  a zinc
  b aluminium
  c boron
  d lead.

8 Name at least three elements that are in the same:
  a group as potassium
  b period as calcium.

INQUIRY

9 Use the periodic table to find at least three elements named after scientists and at least three elements named after countries.

SUCCESS CRITERIA

- I can identify the atomic number of an element.
- I can describe how the periodic table is organised, using terms such as *groups* and *periods*.

3.4 PROPERTIES OF ELEMENTS

LEARNING INTENTION

At the end of this lesson I will be able to predict, using the periodic table, the properties of some common elements.

KEY TERMS

ductile
can be drawn out into a wire

malleable
can be bent

LITERACY LINK

SPEAKING

Choose a section from this lesson to verbally summarise to a partner. Your summary should go for at least 30 seconds. Then swap roles, and listen to your partner's summary of a different section.

NUMERACY LINK

DATA

Determine whether the following elements are solids, liquids or gases at room temperature (25 °C) based on their boiling and melting points:

a BP = 440 °C, MP = 115 °C.
b BP = −196 °C, MP = −210 °C.
c BP = 59 °C, MP = −7 °C.

The elements on the periodic table are divided into three main groups – metals, non-metals and metalloids – based on their physical and chemical properties. By exploring properties of elements, scientists have been able to use them in many ground-breaking discoveries.

1 Metals are malleable and ductile

Metallic elements are found on the left-hand side of the periodic table. Metals are good conductors of heat and electricity. They are shiny, **malleable** (can be bent) and **ductile** (can be drawn out into a wire).

The elements in group 1 of the periodic table are lithium (Li), sodium (Na), potassium (K), rubidium (Rb), caesium (Cs) and francium (Fr). They are known as alkali metals. Alkali metals are shiny and soft and are the most reactive metals. They react with oxygen, water and acids in varying degrees. These metals become more reactive as you go down the group.

The elements in group 2 of the periodic table are beryllium (Be), magnesium (Mg), calcium (Ca), strontium (Sr), barium (Ba) and radium (Ra). They are known as the alkaline earth metals. Similar to those in group 1, group 2 metals are shiny and soft, but aren't quite as reactive as the group 1 metals.

Where are metals located in the periodic table?

2 Transition metals have useful and unique properties

In the middle of the periodic table, the elements in groups 3–12 are called the transition metals. Transition metals have a range of physical and chemical properties. They can form many different compounds with

Figure 3.9 Metals are shiny, malleable, ductile and good conductors of heat and electricity.

other elements. Many of these compounds are brightly coloured.

Some transition metals are reactive. For example, iron (Fe) forms rust with oxygen and water. Others are good conductors of electricity and heat. For example, copper (Cu) is used extensively in electrical equipment, such as wiring and motors. Copper is also used in roofing and plumbing and has many important industrial uses. Other transition metals are not good conductors or very reactive. For example, although gold (Au) is a good conductor, it's one of the least reactive metals. Gold has many uses, including in jewellery.

Where are transition metals located in the periodic table?

3 Non-metals include the halogens and noble gases

On the right-hand side of the periodic table are the non-metals. The two groups of non-metals are the elements of group 17, the halogens, and group 18, the noble gases.

The halogens are the most reactive group of non-metals. The halogens are fluorine (F), chlorine (Cl), bromine (Br), iodine (I) and astatine (At). Halogens react with metals to form ionic compounds. They also react with themselves to form covalent compounds.

The six noble gases in group 18 are helium (He), neon (Ne), argon (Ar), krypton (Kr), xenon (Xe) and radon (Rn). The noble gases are also known as the inert gases because they are very unreactive.

Hydrogen is also a non-metal but it is often put into group 1 with the metals because it has one electron in its valence shell, like the elements of group 1.

Figure 3.10 Non-metals are dull and brittle, and don't conduct electricity or heat.

Where are non-metals located in the periodic table?

4 Metalloids are semimetals

On the periodic table, the metals and the non-metals are separated by a group of six elements – boron (B), silicon (Si), germanium (Ge), arsenic (As), antimony (Sb) and tellurium (Te). Polonium (Po) and astatine (At) are also sometimes included in this group.

These elements have a combination of the properties of both metals and non-metals and so are known as metalloids. Metalloids are usually brittle, sometimes shiny solids. Most metalloids have important applications in industry. For example, silicon is a semiconductor of electricity – it conducts electricity under certain conditions. For this reason, it is used in electronics.

Where are metalloids located on the periodic table?

INVESTIGATION 3.4A

Physical properties of elements

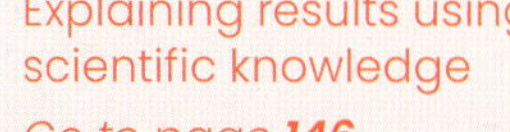

KEY SKILL
Explaining results using scientific knowledge

▶ Go to page **146**

INVESTIGATION 3.4B

Chemical properties of elements

KEY SKILL
Writing a research question

▶ Go to page **147**

CHECKPOINT 3.4

1. a Where on the periodic table are metals located?
   b Identify some properties of metals.
2. How does the reactivity of metals vary as you move down a group or across a period?
3. Why is hydrogen, a non-metal, placed in group 1 with metals?
4. a Where on the periodic table are non-metals located?
   b Identify some properties of non-metals.
5. Sketch the periodic table, showing where to find the noble gases, alkali metals, alkaline earth metals, halogens and transition metals.

RESEARCH

6. Choose an element. Research when and by whom it was discovered, how it is found in nature and its properties and uses.

SUCCESS CRITERIA

- I can identify properties of metals and non-metals.
- I can relate properties to where elements are located on the periodic table.
- I can predict the location of an element on the periodic table from its properties.

3.5 DEVELOPMENT OF THE PERIODIC TABLE

LEARNING INTENTION

At the end of this lesson I will be able to outline how the modern periodic table was developed and refined over time through a process of review by the scientific community.

KEY TERMS

chemical property
a property of a substance that is observed during a chemical reaction

Dalton's atomic theory
the theory that states that all matter is made up of tiny particles

physical property
a property of a substance that can be measured, such as colour, melting and boiling points, density and hardness

LITERACY LINK

READING

Write five questions based on the information in this spread. Give your questions to a partner to answer, while you answer theirs.

NUMERACY LINK

GRAPHING

Construct a timeline showing the development of the periodic table, including names of scientists and their contributions. Ensure the timeline is scaled correctly.

The periodic table developed over thousands of years and is the work of many scientists. The Ancient Greek philosopher Aristotle believed that matter consisted of four elements: water, fire, earth and air. Alchemists developed knowledge of many new elements.

With continued research and scientific evidence, scientists started to group elements according to their physical and chemical similarities. This grouping was important in shaping the current periodic table.

1 Dalton produced an early periodic table of 20 elements

English scientist John Dalton proposed his **atomic theory** in 1808. This theory states that all matter is made of tiny particles called atoms. Dalton produced a table of the mass numbers of 20 elements. He used his own symbols (which were quite difficult to remember) for the elements, and many of the atomic masses he used have been found to be incorrect.

Figure 3.11 Dalton produced an early table of 20 elements.

What did Dalton produce?

2 Döbereiner grouped elements into triads

Scientists continued to expand Dalton's table as they discovered new elements. In 1829, German chemist Johann Döbereiner saw a relationship between the properties of the 55 elements known at the time. He observed trends in the **physical** and **chemical properties** of some elements, known as 'triads'. This grouping of elements was the beginning of the periodic table.

What contribution did Döbereiner make to the periodic table?

3 Newlands arranged elements by mass number

In 1864, English chemist John Newlands was the first person to arrange the 60 known elements in order of increasing mass number. Newlands is known for his law of octaves. When he arranged elements in seven groups of eight, every eighth element seemed to have similar properties. However, some elements did not fit the pattern; for example, Newlands put the metal iron into the same group as the non-metals oxygen and sulphur.

Figure 3.12 Newlands arranged elements in order of increasing mass number.

What was Newlands' contribution to the periodic table?

❹ Mendeleev arranged elements into groups and left gaps

Figure 3.13 Mendeleev arranged elements by atomic mass into vertical columns, or 'families', and predicted the properties of unknown elements.

In 1869, Russian chemist Dmitri Mendeleev arranged the known elements in order of mass number, putting the known 'families' into vertical columns.

He also left gaps in the periodic table, for elements that were not yet discovered. Mendeleev predicted that the unknown elements would have similar chemical properties to those of their 'family'. Later, when the elements were discovered, their properties closely matched Mendeleev's predictions.

The same year, and independently of Mendeleev, German chemist Lothar Meyer constructed a similar table to that of Mendeleev by comparing physical properties of elements.

What was Mendeleev's contribution to the periodic table?

❺ Moseley arranged elements by atomic number

In 1913, English chemist Henry Moseley suggested that it was the atomic number, not the mass number, that was related to the physical and chemical properties. He refined the previous periodic table and came up with a more accurate version with fewer elements missing.

What was Moseley's contribution to the periodic table?

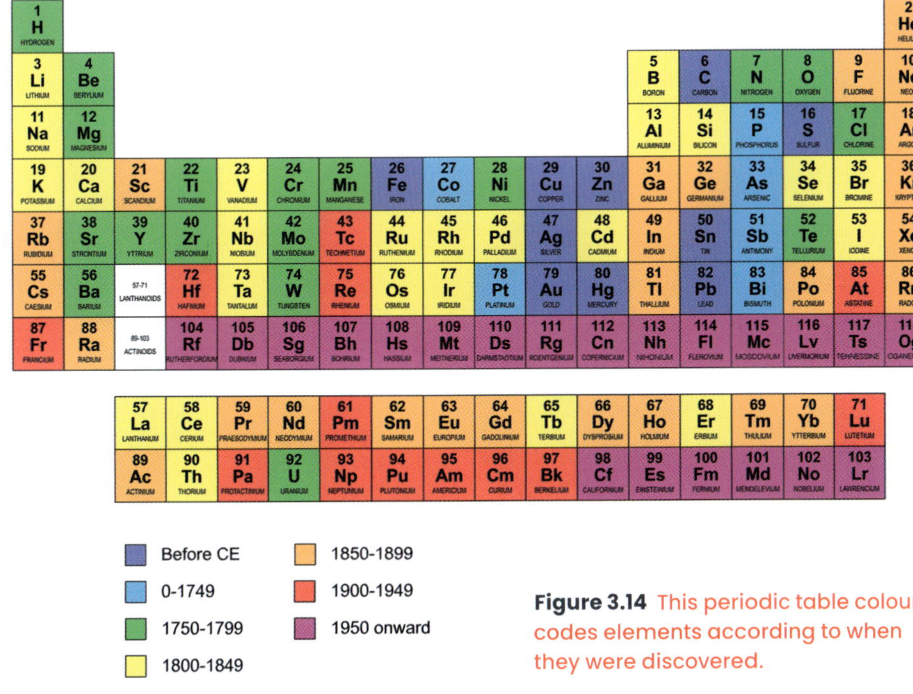

Figure 3.14 This periodic table colour-codes elements according to when they were discovered.

CHECKPOINT 3.5

1 Identify some differences between the early periodic table and the modern periodic table.

2 How did Moseley's suggestion change Mendeleev's periodic table?

3 Describe the law of octaves proposed by Newlands.

4 Over time, more and more elements have been discovered. Suggest some reasons for this.

5 Is there a pattern in the classification of the elements on the periodic table, from the early table made by Dalton to the current one? Explain your response.

6 Research Mendeleev's periodic table. Compare Mendeleev's original periodic table with the current periodic table. What are the similarities and differences?

EXTENSION

7 Draw up a table with three columns headed 'Period of discovery', 'Elements in order of discovery' and 'Total number of elements'. Research all the elements in order of their discovery and complete your table. The periods of discovery could be Ancients, 1600s, 1700s, 1800s, 1900s and 2000s.

SUCCESS CRITERIA

- I can identify the main contributions of scientists in the development of the periodic table.
- I can outline the scientific processes used to develop the periodic table.

VISUAL SUMMARY

The periodic table has evolved over many years and groups together elements based on their physical and chemical similarities.

- All matter is made up of elements.
- An element is made up of identical atoms.
- Every element has a unique symbol.

Dalton proposed the atomic theory and created a table of 20 elements and their atomic numbers and masses.

Döbereiner identified the relationships between the 55 known elements at the time, and grouped elements based on their common traits.

Newlands organised the 60 known elements at the time in order of increasing mass number. However, the rows of his table showed too many elemental differences between the columns.

Mendeleev arranged known elements in order of atomic mass into vertical columns, or 'families'. This allowed Mendeleev to predict the properties of unknown elements.

Moseley refined the periodic table, suggesting that an element's physical and chemical properties were related to the atomic number, not the atomic mass.

Most element symbols are easy to identify (e.g. C for carbon). Some symbols are not as obvious because they originate from the Latin, Greek or German name of the element (e.g. Fe for iron comes from the Latin word *ferrum*.)

The atomic number is the number of protons in the nucleus of an atom.

The mass number is the number of protons and neutrons in the nucleus of an atom.

Elements up to calcium follow the 2, 8, 8, 2 rule. Electrons fill the innermost shells first.

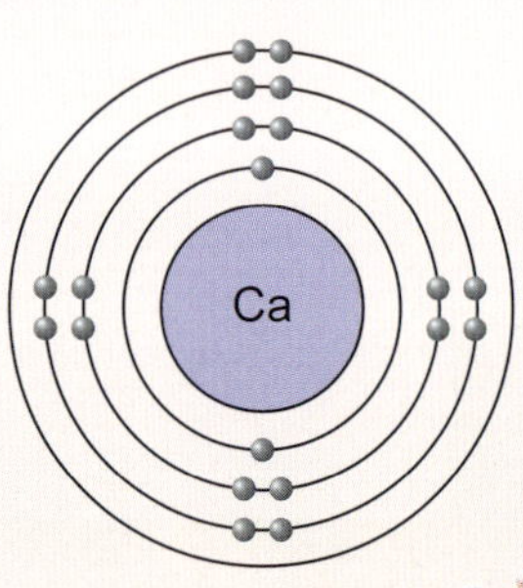

Metals: are good conductors of heat and electricity; are shiny, malleable and ductile

react with oxygen, water and acids (to certain degrees).

Metalloids have properties of both metals and non-metals.

Non-metals:

- can react with metals, or themselves (halogens)
- are not reactive (noble gases)
- do not conduct electricity and heat
- are dull and brittle.

★ FINAL CHALLENGE ★

1. Identify the symbols for the elements chromium, curium, strontium and antimony.
2. Identify the number of protons, neutron and electrons for the elements calcium, aluminium and xenon.

LEVEL UP!
Level 1
50xp

3. This list shows scientists and their contributions to the discovery of the atom. Draw an arrow between the scientist and their contribution.

Mendeleev	Produced a table showing the symbols and atomic masses of 20 elements
Döbereiner	Arranged the 60 known elements in order of increasing atomic mass
Dalton	Arranged the known elements in order of mass number, putting the known 'families' into vertical columns
Newlands	Grouped elements belonging to the same 'family' according to their physical and chemical properties
Moseley	Constructed a similar table to that of Mendeleev by comparing physical properties of elements with atomic mass

LEVEL UP!
Level 2
100xp

4. Use the periodic table to determine the information to complete this table for elements 11–20.

Element	Symbol	Atomic number	Mass number	Number of protons	Number of neutrons	Number of electrons

5. Describe some physical and chemical properties that distinguish metals from non-metals.

6. Why do you think the periodic table is arranged by atomic number and not by mass number? Explain your reasoning.
7. Draw a diagram and state the electron configuration of:
   a. fluorine
   b. sulphur
   c. magnesium.

8. Predict which group the following elements belong to, based on the characteristics described.
   a. Element A: shiny, ductile and very reactive
   b. Element B: good conductor of heat and electricity
   c. Element C: brittle, shiny, solid
9. What can the period and group of an element tell us about its atomic structure?

CHEMICAL REACTIONS

How does the chemical world shape our lives?

Have you ever wondered why some reactions are very slow (for example, the spoiling of milk and corrosion) while others are fast (for example, the combustion reactions that happen in bushfires)? The rate of a reaction is a measure of how quickly reactants are used up and products are formed. If this happens fast, then the rate of reaction is high. If a reaction takes a long time to complete, then the reaction rate is low.

1 FRAYER MODEL

Copy and complete the below chart in your workbook.

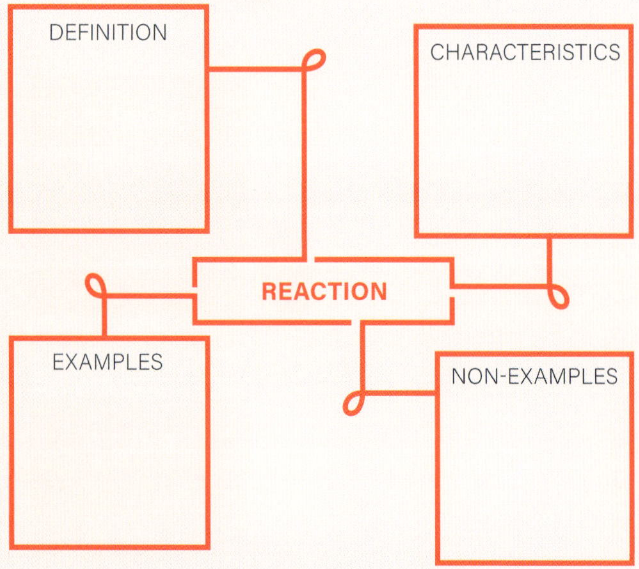

Complete two additional charts for the key terms *Compound* and *Rate*.

2 LEARNING LINKS

Brainstorm everything you already know about chemical reactions.

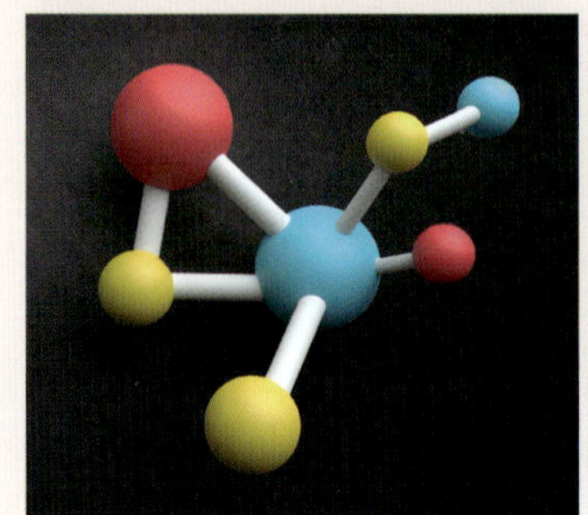

3 SEE-KNOW-WONDER

List three things you can SEE, three things you KNOW and three things you WONDER about this image.

4 CRITICAL + CREATIVE THINKING

WHAT IF ... combustion reactions (burning things) was banned in Australia?

THE VARIATIONS: How many different ways can you use chemical reactions every day?

THE COMBINATION: List two features of a catalyst and two features of a submarine. Now combine those attributes to create a new object.

5 THE WORST CHEMICAL DISASTER!

Huge chemical disasters are rare, but they still happen all over the world. Perhaps the worst such disaster was a major gas leak at a pesticide plant in Bhopal, India, in 1984. Almost 10 000 people died as a result of poisoning from the leak, and more than 500 000 people were injured. The site of the leak is still contaminated, and people living in Bhopal still become ill from exposure to toxic chemicals.

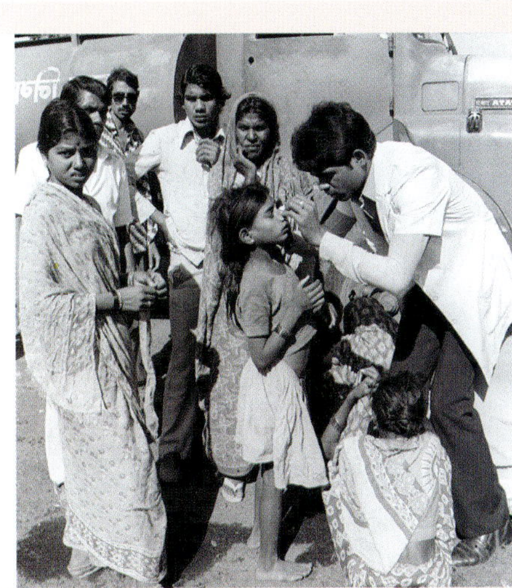

4.1 ATOMS AND MATTER

LEARNING INTENTION

At the end of this lesson I will be able to recall that all matter is made up of atoms and has mass.

KEY TERMS

compound
a substance made up of two or more types of atoms bonded together in fixed ratios

isotope
an atom of a particular element with a different number of neutrons

mass
the amount of matter in an object

matter
a physical substance that has mass and takes up space

pure substance
a substance that contains only one type of particle

LITERACY LINK

WRITING

Design and write a numbered method for an investigation that aims to prove that air has mass. (Hint: You can use a balance.)

NUMERACY LINK

CALCULATION

The mass of a hydrogen atom is 1.67×10^{-27} kg. Calculate the mass of a deuterium and a tritium atom.

Everything in the universe is made up of matter. Matter is anything living or non-living that occupies space and has mass. It can be in the form of a solid, liquid, gas or even plasma. You, water, books, tables, the Sun, Earth, air, cars and trees are all made up of matter.

Matter is made of atoms, and those atoms determine a material's physical and chemical properties. An atom is the smallest basic unit of matter, so small that you cannot see it even with very strong optical microscopes. So, in order to understand the world around you, you need to visualise what an atom is and how it behaves.

1 Matter is made up of different types of atoms

Everything is made up of a combination of different types of atoms. Atoms of identical size, mass, and similar properties are called elements. These are listed in the periodic table of elements.

Atoms consist of a positive nucleus containing protons and neutrons surrounded by negative electrons orbiting the nucleus. The mass of the atom is concentrated inside its nucleus and is the sum of the mass of the protons and neutrons.

What are the main components of an atom?

Figure 4.1 Matter is made up of atoms. Atoms contain protons and neutrons inside the nucleus and electrons in energy levels outside the nucleus.

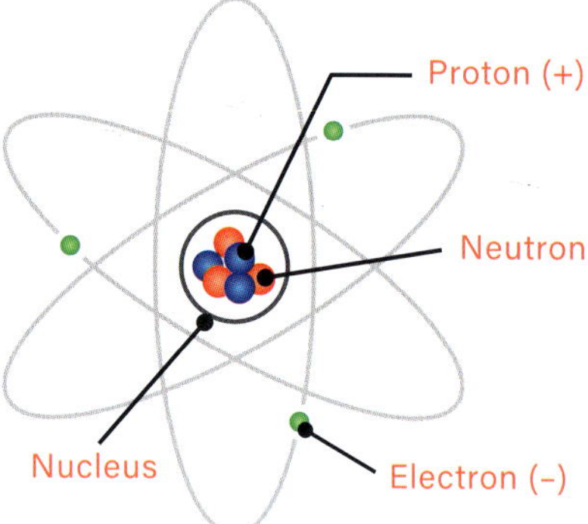

2 Pure substances can be elements or compounds

A **pure substance** is a substance that only contains one type of particle. Pure substances can either be elements or **compounds**.

An element is a substance that is made up of only one type of atom. For example, a block of sodium is made up of billions of identical sodium atoms. All of the known elements are found on the periodic table.

A compound is a substance made up of more than one type of atom held together in fixed ratios by chemical bonds. Pure water (H_2O) is made up of only water particles, and each particle contains two hydrogen atoms and one oxygen atom, chemically bound together. Water is both a compound (it contains two different types of atoms: hydrogen and oxygen) and a pure substance (it is only made of water particles). Other compounds include carbon dioxide (CO_2), glucose ($C_6H_{12}O_6$) and salt (NaCl).

What is a compound?

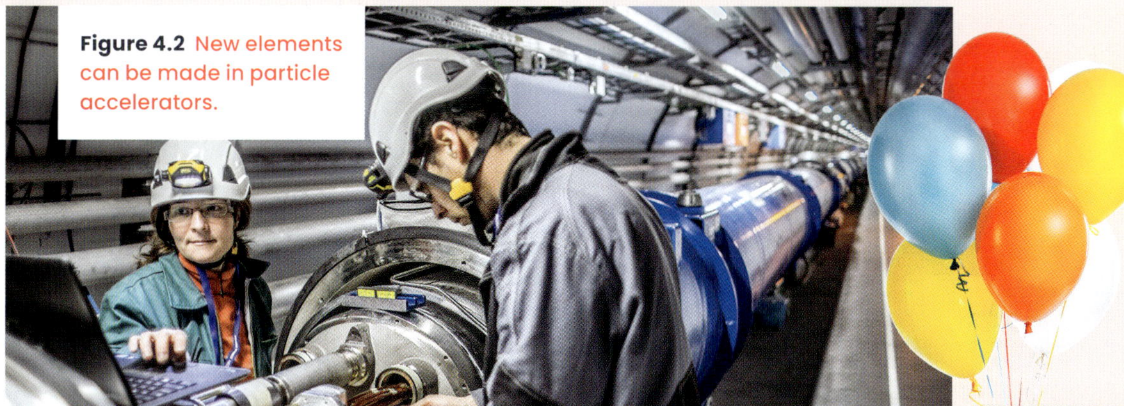

Figure 4.2 New elements can be made in particle accelerators.

Figure 4.3 A helium balloon has a small amount of mass in a large volume.

3 Matter has mass

Mass is the amount of **matter** in an object. A small object such as a copper bell might have a lot of mass. A large object such as a balloon filled with helium might have very little mass.

It's important to remember that mass is not weight. Mass is a measure of the matter in an object, while weight is a measure of gravity's pull on an object. An object's mass is constant; you can only change its mass by changing the object, such as breaking off a piece of it.

What is the difference between mass and weight?

4 Isotopes are different forms of the same element

Some elements exist in different forms called **isotopes**. Isotopes have different numbers of neutrons in the nuclei of their atoms. This means that isotopes have different mass numbers and some different properties. For example, many isotopes are radioactive. However, all isotopes of an element have the same atomic number and occupy the same space on the periodic table.

The isotopes of hydrogen (Figure 4.3) have one proton but different numbers of neutrons. They also have different names.

What is an isotope?

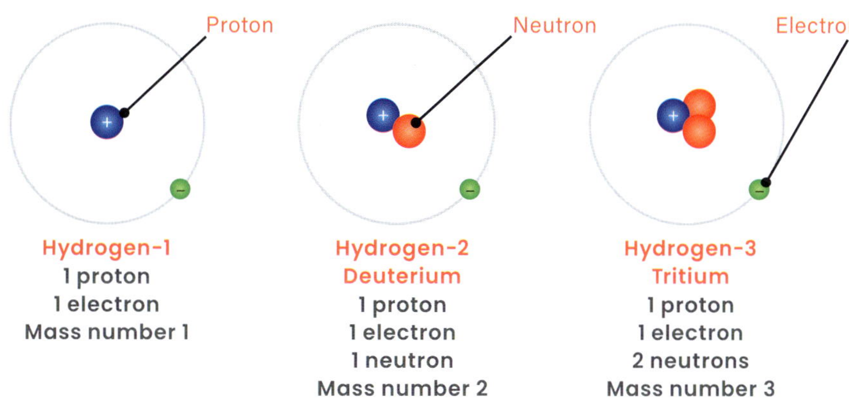

Figure 4.4 The isotopes of hydrogen are called deuterium and tritium.

CHECKPOINT 4.1

1. Explain what matter is.
2. Describe the difference between an element and a compound.
3. A compound can be a pure substance, even though it's made of different types of atoms. Explain.
4. The Moon has much weaker gravity than Earth. How would an astronaut's mass and weight differ on the Moon compared to on Earth?
5. Draw the atomic structure of an atom and label the nucleus, protons, electron and neutrons.
6. What is an isotope, and how do isotopes of the same element differ?
7. What do you think the number in an isotope name (e.g. Hydrogen-2) means? *Hint: see Figure 4.4.*
8. State the number of protons, electrons and neutrons in the carbon isotopes C-12, C-13 and C-14.

RESEARCH

9. Research to find out why scientists made technetium.

SUCCESS CRITERIA

- I can explain what matter is.
- I can describe the structure of an atom.
- I can explain what an isotope is.

4.2 BONDS BETWEEN ATOMS

LEARNING INTENTION

At the end of this lesson I will be able to classify compounds into groups based on their bonding type.

KEY TERMS

covalent bond
a bond in which two atoms share one or more pairs of electrons

ion
an atom with unequal numbers of protons and electrons; a charged atom

ionic bond
a bond in which some atoms gain and some lose electrons, becoming ions

lattice
an interlaced structure or pattern

metallic bond
a bond in which free electrons move around metal ions

molecule
the smallest unit of a covalent compound that can take part in a chemical reaction

valency
an atom's power to combine with other atoms

NUMERACY LINK

GRAPHING

The mass number of nitrogen (N) is 14 and hydrogen (H) is 1. Ammonia (NH_3), therefore has a mass of 17. Calculate the masses of the molecules in Table 4.1 and present your results as a bar chart.

All matter is made up of atoms, but atoms are incredibly small particles. Any piece of matter that you can see, even something as small as a grain of sand, contains millions of atoms.

In order for atoms to stick together to make a compound, they need to form bonds with other atoms. These bonds are formed by the atoms' electrons, and there are three different types of bonds that can form.

1 Covalent compounds are bound together by shared electrons

Electrons move around the nucleus of atoms in different orbits or 'shells'. The electrons in the outermost shell, called the valence shell, move around to bond atoms together. An atom's ability to combine with other atoms is called its **valency**.

When two or more non-metallic atoms bond, they share electrons in their valence shells. The shared electrons orbit the nuclei of all the bonded atoms, producing a **molecule**. This kind of bond is called a **covalent bond**, and compounds formed this way are called covalent compounds.

While a few non-metallic elements, such as helium, naturally occur as single atoms, most non-metallic elements appear in nature as molecules containing two or more atoms.

Table 4.1 Some common molecules and their formulas

Name	Formula
Ammonia	NH_3
Carbon dioxide	CO_2
Chlorine	Cl_2
Glucose	$C_6H_{12}O_6$
Hydrogen	H_2
Ozone	O_3
Water	H_2O

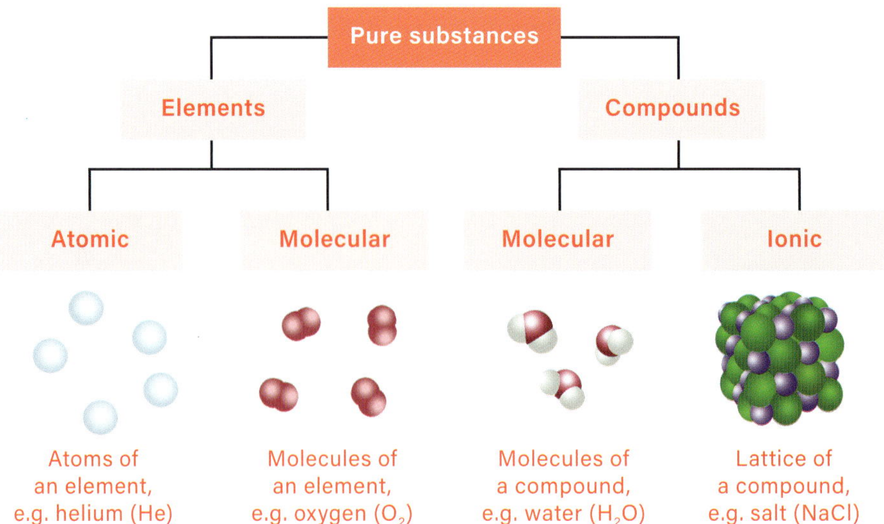

Figure 4.5 Pure substances can be classified as elements or compounds.

A molecule is the smallest unit of a covalent compound that can take part in a chemical reaction. Some examples of molecules are water, oxygen, nitrogen and glucose. Some more molecules and their formulas are listed in Table 4.1.

How are covalent compounds formed?

2 Ionic compounds are bound together by electron loss and gain

Some atoms gain or lose electrons when they bond, creating **ions**. Ions are still atoms, but they are atoms with an electrostatic charge. When atoms lose electrons, they form positively charged ions; when they gain electrons, they form negatively charged ions.

When a metal atom bonds with a non-metal atom, the metal loses electrons and the non-metal gains them. This creates an ionic compound, such as sodium chloride (NaCl), made up of positive (metal) and negative (non-metal) ions. Rather than individual molecules, the ions arrange themselves into a **lattice**. The **ionic bonds** in an ionic compound are very strong, because of the electrostatic attraction between the positive and negative ions.

Ions are shown by writing a + or – sign at the upper-right of the formula, such as Na^+ and Cl^-. Some ions are formed by gaining or losing more than one electron. We indicate this by writing a number in front of the + or –, such as O^{2-} or Hg^{2+}.

How are ionic compounds formed?

3 Metal atoms are bound together by free electrons

In metallic bonding, all of the metal atoms lose electrons, rather than gain or share them. These lost electrons form a 'sea' of free electrons that surrounds the positive metal ions. The metal ions are closely packed together and are held in place by their attraction to the sea of electrons.

Metallic bonds tend to be weaker than ionic bonds, but the large numbers of free electrons make metallic compounds good conductors of heat and electricity.

How are metals formed?

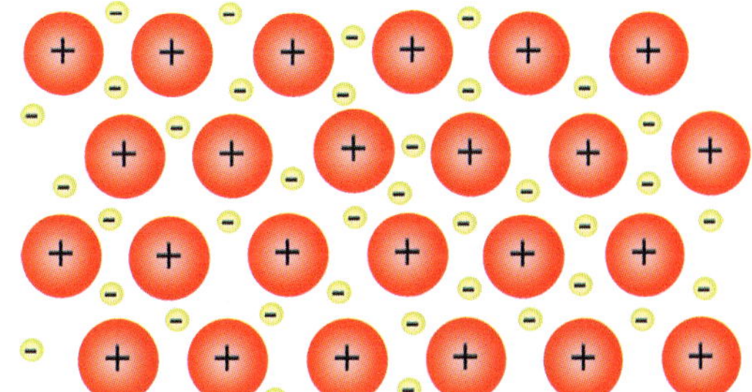

Figure 4.6 Free electrons move easily around metal ions in a metal.

CHECKPOINT 4.2

1 Explain what a compound is.

2 Describe the difference between covalent, ionic and metallic bonding in terms of electrons.

3 How are ions formed?

4 Glucose has the formula $C_6H_{12}O_6$. How many atoms of carbon, hydrogen and oxygen are there in one molecule of glucose?

5 How are covalent bonds formed?

6 In terms of metals and non-metals, state the types of atoms that will form:
a covalent bonds
b ionic bonds
c metallic bonds.

7 Do metals form positive or negative ions when they lose electrons? Explain.

8 Explain the following statement.
Metals are never compounds, ionic substances are always compounds, and covalent substances can be compounds.

CONNECTING IDEAS

9 For a substance to conduct electricity, it must contain free-moving charged particles. Explain why metals are able to conduct electricity, but covalent and solid ionic substances are not.

SUCCESS CRITERIA

- I can explain what a compound is.
- I can classify compounds into groups based on their type of bonding.

4.3 IDENTIFYING CHEMICAL COMPOUNDS

LEARNING INTENTION

At the end of this lesson I will be able to identify a range of compounds using their names and chemical formulas.

KEY TERMS

chemical formula
chemical symbols showing the ratio of elements to one another

polyatomic ion
two or more ions bonded together and acting as a single charged unit

prefix
a word or number placed before another word

LITERACY LINK

VOCABULARY

The prefix *tri-* means *three*, as in phosphorous trichloride, PCl_3. Write a list of five other words that begin with *tri-* and state their relationship to the number three.

NUMERACY LINK

CALCULATION

If the formula for a particular ionic compound is X_2O_3. Determine the charge (valency) of X.

A chemical formula is a shorthand way of writing the name of an element or compound. It tells you the number and type of atoms in a compound. You can then use these formulas to write chemical equations that model reactions.

1 There are rules for naming covalent compounds

While many compounds have simple, everyday names – water, vinegar, bleach – they also have scientific names. These two-word names are based on their **chemical formula**.

To make sure formulas can be understood in every language, scientists follow four rules when naming compounds.

1 Determine whether the compound is ionic or covalent. If it contains a metal, it is ionic. If it only contains non-metals, it is covalent.
2 If it is covalent, name the element that comes first in the formula.
3 Name the second element, making sure it ends with 'ide'; for example, carbon fluoride (CF), hydrogen chloride (HCl).
4 If there is more than one of a particular type of atom in the formula, a **prefix** can be used to indicate the number of that type of atom; for example, diphosphorus pentoxide (P_2O_5), carbon dioxide (CO_2).

For example, the chemical formula for water is H_2O. Following our naming rules, (1) we can see that it contains no metals and so is covalent, (2) the hydrogen comes first, (3) the oxygen becomes oxide, and (4) there are two hydrogens, and so these need the prefix 'di'. So, water's scientific name is dihydrogen monoxide.

Table 4.2 Number of atoms and the prefixes to use

Number of atoms	1	2	3	4	5	6	7	8
Prefix	mono-	di-	tri-	tetra-	penta-	hexa-	hepta-	octa-

Why do scientists follow rules when naming compounds?

Figure 4.7 Water, vinegar, baking soda and bleach are the common names for, respectively, dihydrogen monoxide, ethanoic acid, sodium bicarbonate and sodium hypochlorite.

❷ Ionic compounds require additional rules

Although the basic rules for naming compounds are simple, it can become more complicated when naming ionic compounds.

The first change is that the metal atom in the compound is always named first. Some ionic compounds are made up of more than two elements. In this case, there is a group of elements bonded together into a **polyatomic ion**. When this happens, the metal is named first, then the polyatomic ion, which has its own name; for example, sodium sulfate (Na_2SO_4), copper hydroxide ($Cu(OH)_2$). (These names don't always follow the 'ide' rule.)

Table 4.3 Common polyatomic ions

Name	Formula
Ammonium	NH_4^+
Carbonate	CO_3^{2-}
Hydroxide	OH^-
Nitrate	NO_3^-
Nitrite	NO_2^-
Phosphate	PO_4^{3-}
Sulfate	SO_4^{2-}
Hydrogen sulfate	HSO_4^-
Sulfite	SO_3^{2-}

To determine the chemical formula for an ionic compound, you can use the swap-and-drop method.

Step 1 Write the ions with the positive ion on the left and the negative ion on the right.

Step 2 Swap the charge number on the left-hand side to become a subscript on the right-hand side.

Step 3 Swap the charge number on the right-hand side to become a subscript on the left-hand side.

Step 4 Simplify, using brackets and the lowest common factor.

For example, calcium carbonate is a compound in which calcium (Ca^{2+}) is bonded with carbonate (CO_3^{2-}). To determine the chemical formula of calcium carbonate, using the swap-and-drop method:

Step 1
Write the ions:

Ca^{2+} and CO_3^{2-}

Step 2
Swap left-hand charge:

Ca and $(CO_3^{2-})_2$

Step 3
Swap right-hand charge:

$Ca_2 \quad (CO_3)_2$

Step 4
Simplify:

$CaCO_3$

Some metals may have different ionic forms. Add Roman numerals after the name of the metal to show the ionic form. For example, iron can exist as Fe^{2+} and Fe^{3+}, which form different compounds – iron(II) chloride is $FeCl_2$, while iron(III) chloride is $FeCl_3$.

What is a polyatomic ion?

Figure 4.8 Iron can have different ionic forms: iron(II) and iron(III). Both are soluble in water but they form different coloured solutions.

CHECKPOINT 4.3

1 Explain why chemical formulas are used instead of chemical names.

2 When naming a compound, which element should you write first?

3 Name these compounds.
   a FeS
   b KBr
   c $PbSO_4$
   d $CuCO_3$
   e $Ca(OH)_2$

4 Write the chemical formulas for these compounds.
   a potassium nitrate
   b barium sulfate
   c sodium oxide
   d ammonium chloride

5 Explain what a polyatomic ion is.

INQUIRY

6 Try to find out where acetylsalicylic acid, calcium hydroxide, calcium carbonate and monosodium glutamate are found in your home.

SUCCESS CRITERIA

- I can identify a range of compounds using their names and chemical formulas.
- I can use rules to name chemical compounds.

4.4 WRITING CHEMICAL EQUATIONS

LEARNING INTENTION

At the end of this lesson I will be able to use word or symbol equations to represent chemical reactions.

KEY TERMS

coefficient
a number placed before the chemical in a formula or chemical equation

conservation of mass
the law that mass cannot be created or destroyed

product
a substance formed in a chemical reaction

reactant
a substance that starts a chemical reaction

LITERACY LINK

WRITING

Create a flow chart to demonstrate the steps you would use to balance a chemical reaction.

NUMERACY LINK

UNITS

Refer to the equation $N_2 + 3H_2 \rightarrow 2NH_3$

If there are 3 molecules of N_2, how many molecules of H_2 are required, and how many molecules of NH_3 are produced?

One of the unbreakable laws of the universe is that matter cannot be created or destroyed; this is the law of **conservation of mass**. One of the effects of this law is that the amount of matter at the start of a chemical reaction is always the same as the amount of matter left at the end. The atoms and molecules may be different, but the total mass never changes.

An important application of this law is that it lets us model a chemical reaction by writing a chemical equation, which is very similar to an equation in maths. A chemical equation shows what atoms are involved before the reaction and how they recombine afterwards.

1 Start by listing the reactants and the products

The first step in writing a chemical equation is to list all of the substances involved. The **reactants** are the substances that you have before the reaction takes place, while the **products** are the substances you will have after the reaction has finished.

Write the reactants on the left side of an arrow, and the products on the right side, using plus signs to separate each material on each side. Although we use formulas in the equation, it's often easiest to start by writing the names of the materials. This is known as a word equation. You can then write the formulas underneath the words.

A simple chemical reaction is the one in which nitrogen and hydrogen combine to form ammonia, which is an important component of fertilisers. This takes place at high temperatures. The chemical equation for this reaction would start off looking like this:

nitrogen	+	hydrogen	→	ammonia
N_2	+	H_2	→	NH_3

What is the difference between a reactant and a product?

Figure 4.9 Producing ammonia fertiliser from nitrogen in the air allows farmers to grow more crops.

❷ Balance the number of atoms on each side

The law of conservation of mass says that matter cannot be created or destroyed. Whenever it looks as though matter has been destroyed, it has actually been lost in the form of gas.

In a chemical reaction, the atoms of the reactants rearrange to form the molecules of the products. Because no new atoms can be created or destroyed, this means that the number of atoms of each element must be the same on both sides of the equation. If the numbers do not match, you need to balance the equation by adding more molecules to one or both sides. Do this by adding **coefficients** (numbers) in front of the reactants or products to show that there are more molecules of that substance.

Consider the ammonia formation equation:

N_2 + H_2 → NH_3

Step 1:

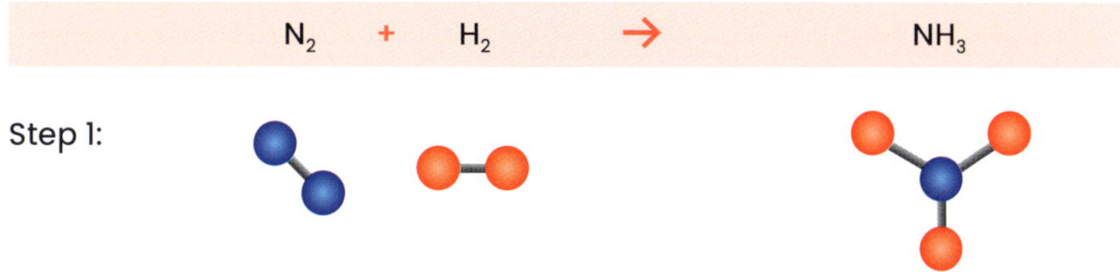

This equation is unbalanced because there is not the same number of nitrogen atoms and hydrogen atoms on each side.

First, balance the number of nitrogen atoms by adding a 2 in front of the NH_3 on the right-hand side:

N_2 + H_2 → $2NH_3$

Step 2:

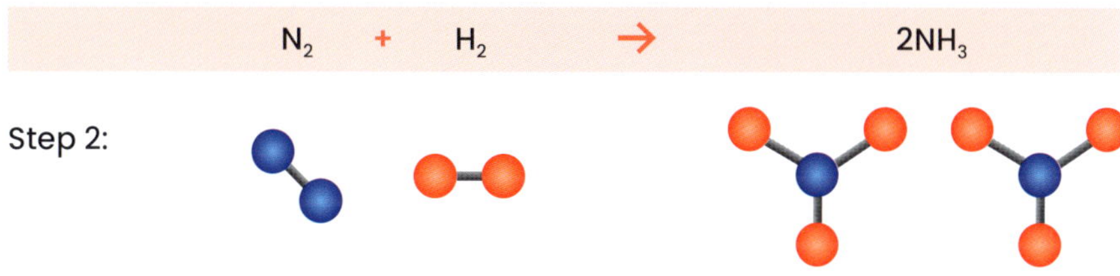

There are now two nitrogen atoms on each side of the equation. However, there are now two hydrogen atoms on the left-hand side and 2 × 3 = 6 hydrogen atoms on the right-hand side. To balance it, put a 3 in front of the H_2 on the left-hand side:

N_2 + $3H_2$ → $2NH_3$

Step 3:

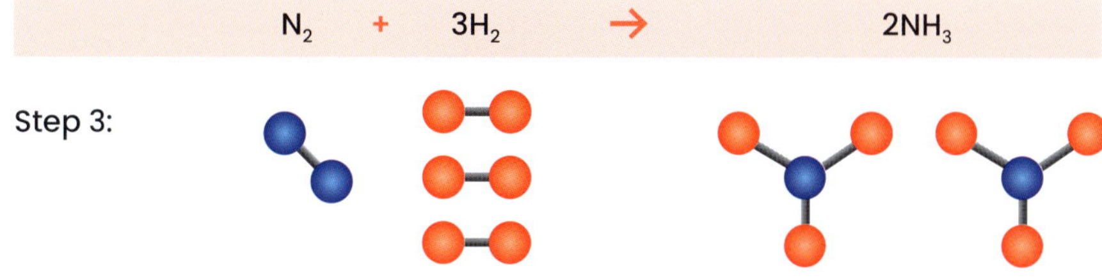

The equation is now balanced, with the same number of atoms on each side. The equation also shows us that one nitrogen molecule and three hydrogen molecules react to form two ammonia molecules.

Why do we need to balance the atoms on each side of a chemical equation?

CHECKPOINT 4.4

1 Explain what the law of conservation of mass is and how it applies to chemical reactions.

2 Explain what the reactants and products are in a chemical reaction.

3 Consider this chemical reaction

$Fe_2O_3 + C \rightarrow Fe + CO_2$

Identify the:
- a products
- b reactants
- c number of iron atoms in each of the reactants and products
- d number of oxygen atoms in each of the reactants and products.

4 Is the equation in question 3 balanced? Why or why not?

5 Balance these equations.
- a $Fe_2O_3 + C \rightarrow Fe + CO_2$
- b $P_4 + O_2 \rightarrow P_2O_5$
- c $NaBr + Cl_2 \rightarrow NaCl + Br_2$

EXTENSION

6 Balance these (more difficult) equations.
- a $C_2H_6 + O_2 \rightarrow CO_2 + H_2O$
- b $Al_2O_3 + Fe \rightarrow Fe_3O_4 + Al$
- c $NH_3 + CuO \rightarrow Cu + N_2 + H_2O$

SUCCESS CRITERIA

- I can balance chemical equations by making sure that the number and types of atoms are the same on both sides of the arrow.
- I can construct word equations from observations and written descriptions of a range of chemical reactions.
- I can explain the law of conservation of mass and how it applies to chemical reactions.

4.5 TEMPERATURE AND CONCENTRATION

LEARNING INTENTION

At the end of this lesson I will be able to investigate the effects of temperature and concentration on the rate of chemical reactions.

KEY TERMS

concentration
the number of particles in a given volume

kinetic energy
energy of motion

LITERACY LINK

LISTENING

Verbally summarise one section from this lesson to a partner, but deliberately make a mistake. Your partner has to try to identify your mistake. Then swap roles and try to find your partner's mistake.

NUMERACY LINK

GRAPHING

Observe the following data:

Concentration of reactant (g/L)	Time of reaction (s)
0	0
2	8
4	14
6	20
8	24
10	24

Draw a line graph and describe the pattern of results.

Reactions can be fast, such as an explosion or firing a missile. Reactions can also be slow, such as a ship rusting or rocks weathering, which take many years. The rate of a chemical reaction is how fast the reaction can be completed; that is, the speed at which reactants are converted into products. Many factors influence the rate of reaction, including temperature, concentration, the presence of a catalyst and the surface area involved.

Figure 4.10 **(a)** Wood burning is a fast chemical reaction, which may be completed in a few hours. **(b)** Iron rusting is a slow chemical reaction, which can take many years.

1 Many factors can affect reaction rate

Every chemical reaction makes products from the reactants at a certain rate, but this rate is not fixed. Many different factors affect reaction rate, including:

- concentration
- temperature
- pressure
- surface area
- catalysts.

It's possible to adjust many of these factors. In some cases, you might want to reduce the reaction rate to make the reaction safer to handle. Or you might want to increase the rate of a reaction so that products are made more quickly. This is very common in the chemical industry, where it is important to reduce costs and maximise profits.

What are three factors that can affect reaction rate?

❷ Raising temperature increases reaction rate

Molecules must collide in order to react. Increasing the temperature of a system increases the average **kinetic energy** of the particles in the system. As kinetic energy increases, the particles move faster, so they collide more frequently and with greater energy.

The reaction rate of nearly all reactions increases with increasing temperature. This means that the hotter the reactants, the faster the reaction will be. Likewise, the cooler the reactants, the slower the reaction.

We change temperature to adjust reaction rates every day. You boil water before making a cup of tea because the tea steeps faster in hot water. If you make a cup of tea from room-temperature water, it will take much longer. Similarly, we keep food in the refrigerator to slow down the chemical reactions that cause it to spoil over time.

Figure 4.11 The reactions in glow sticks give off light. When a glow stick is placed in warm water, it glows more brightly because the reactions happen faster at higher temperatures.

How does changing the temperature affect reaction rates?

❸ Concentrated substances react more quickly

The **concentration** of a substance is a measure of how many particles are within the volume of a liquid solution or a gas's container. Increasing the concentration of the reactants increases the number of particles within a given volume. The more crowded the particles are, the more collisions occur. This increase in the frequency of collisions leads to an increase in the rate of the reaction.

Another way to adjust concentration for a gas is by changing the pressure. The pressure of a gas increases when the volume of the container holding the gas decreases. This means that there are now the same number of particles in a smaller volume, leading to more collisions between particles.

How does changing the concentration affect reaction rates?

Figure 4.12 Concentrated sulphuric acid reacts dramatically with sucrose (table sugar).

INVESTIGATION 4.5A

The effect of concentration on the rate of a reaction

KEY SKILL
Identifying and managing relevant risks

▶ *Go to page* ***148***

INVESTIGATION 4.5B

The effect of temperature on the rate of a reaction

KEY SKILL
Representing data to identify patterns and trends

▶ *Go to page* ***150***

CHECKPOINT 4.5

1 Explain why temperature can increase reaction rate.
2 Outline why collisions between the reacting particles are important for a chemical reaction to take place.
3 Will increasing the concentration of one reactant increase the rate of the reaction? Explain your answer.
4 Is it always a good thing to increase the reaction rate? Explain your answer.
5 Explain how the rate of a gaseous reaction is affected when the volume of the container decreases.

CRITICAL AND CREATIVE THINKING

6 Design a machine that could be used to vary the rate of a chemical reaction. Make sure to use more than one method of changing rates in your design.

SUCCESS CRITERIA

- I can describe the effect of temperature on the rate of chemical reactions.
- I can describe the effect of concentration on the rate of chemical reactions.

4.6 SURFACE AREA AND CATALYSTS

LEARNING INTENTION

At the end of this lesson I will be able to investigate the effects of surface area and catalysts on the rate of chemical reactions.

KEY TERMS

activation energy
the energy required to start a reaction

catalyst
substance that increases the rate of a chemical reaction without being used up

enzyme
a biological catalyst that increases the rate of reactions in cells

surface area
the area of the outermost layer of an object

LITERACY LINK

VOCABULARY

The word *enzyme* is often used in advertising. Think of three other scientific words that are often used in advertising, and determine their relationship (if any!) to the product.

NUMERACY LINK

CALCULATION

Determine the surface area of a cube with 1 cm sides. Then calculate the total surface area of the same cube cut into eight pieces. Does the large cube or the smaller cubes combined have a larger surface area?

Temperature and concentration are relatively easy factors to control. We adjust these factors every day in our kitchens. Other factors are harder to control, and may require special equipment or materials. These factors are often adjusted in industrial or manufacturing processes, although we can adjust them in smaller ways in our own lives.

1 Increasing surface area increases reaction rate

Surface area is the exposed area of a solid substance. If you increase the surface area of a substance, there are more particles at the surface that can react with another substance, so the reaction rate increases.

Consider the reaction of zinc metal with an acid (Figure 4.13). If the zinc is present as a large cube, the total surface area of the cube is the sum of all the faces of the cube. If you break the zinc cube into many smaller pieces, the total surface area greatly increases, and more area is exposed to react with the acid. If you crush zinc into powder, its surface area increases even further. This form of zinc reacts the fastest.

An everyday example is that small pieces of food cook faster. These pieces of food have a large surface area, compared to their volume, exposed to heat. Medicinal tablets are another example. When you take tablets or pills, different tablets dissolve at different rates depending on their surface areas.

Why does increasing the surface area increase the rate of a chemical reaction?

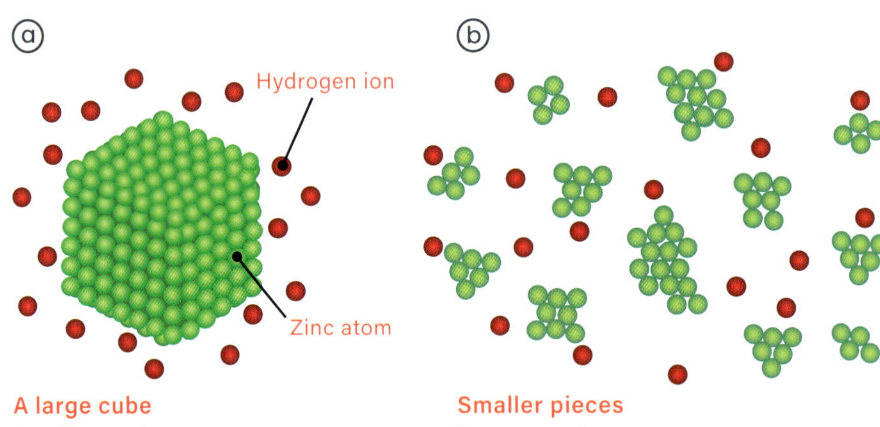

Figure 4.13 **(a)** Hydrogen ions from the acid can collide only with the outer layer of zinc atoms in the large cube of zinc. The reaction rate is slow. **(b)** When the zinc block is broken into smaller pieces, more surface is exposed for the hydrogen ions to collide with and the rate of the reaction increases.

❷ Catalysts increase the rate of reaction

A **catalyst** is a substance that increases the rate of a chemical reaction without being used up. It enables the reaction to occur at a lower temperature.

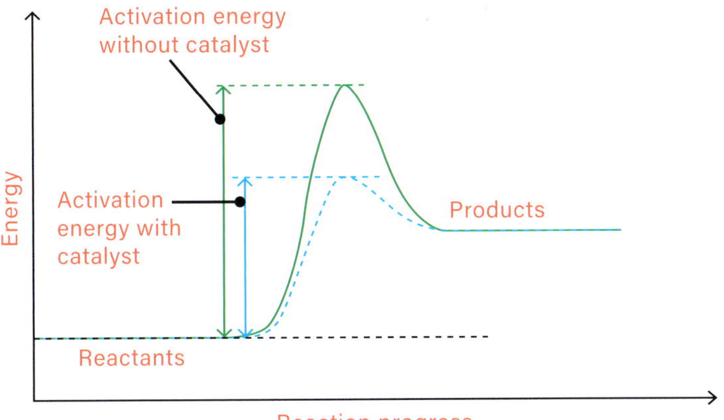

Figure 4.14 The chemical reaction that is catalysed requires less activation energy.

Catalysts lower the **activation energy** of a reaction, which is the energy required to start a reaction. Catalysts are specific, which means that each catalyst increases the rate of some reactions but not others.

Catalysts are important in industry because they reduce reaction times and enable chemical processes to take place at lower temperatures and pressures. So they reduce the cost of production of valuable products.

Enzymes are biological catalysts. They are proteins that help speed up chemical reactions in the body.

Here are some common uses of catalysts.

- Catalytic converters in cars contain platinum, which catalyses the conversion of toxic carbon monoxide into carbon dioxide.
- Biological catalysts in detergents remove protein stains, such as egg yolk, that are otherwise hard to dislodge.
- Catalysts in the paper industry break down paper pulp to produce smooth paper for magazines.
- Catalysts turn milk into yoghurt and petroleum into plastic milk jugs and bicycle helmets.

Why are catalysts important in our lives?

Figure 4.15 Saliva contains enzymes that start the chemical reactions of digestion.

INVESTIGATION 4.6A

The effect of a catalyst on the rate of a reaction

KEY SKILL
Drawing conclusions consistent with evidence

▶ *Go to page* ***152***

INVESTIGATION 4.6B

The effect of surface area on the rate of a reaction

KEY SKILL
Identifying the variables and formulating a hypothesis

▶ *Go to page* ***154***

CHECKPOINT 4.6

1 Explain how surface area can influence reaction rate.

2 Describe activation energy.

3 Explain the significance of the number of collisions to the rate of a chemical reaction.

4 Explain what a catalyst is and give an example of a catalyst.

5 Explain how enzymes are a type of catalyst.

6 True or false? Crushing a substance increases its surface area.

RESEARCH

7 Research catalysts in industry. Identify a catalyst and state where it is used. How does this catalyst help improve the rate of this reaction? Write the word and formula equations for the reactions that take place. What would have happened if this catalyst hadn't been used?

SUCCESS CRITERIA

- I can describe the effect of increasing surface area on the rate of chemical reactions.
- I can describe the effect of a catalyst on the rate of chemical reactions.

4.7 CARBON CAPTURE TECHNOLOGY

LEARNING INTENTION

At the end of this lesson I will be able to analyse how social, ethical and environmental considerations can influence decisions about scientific research.

KEY TERMS

carbon capture
the process of trapping carbon dioxide at its emission source so that it does not enter the atmosphere

implication
a likely outcome or consequence of an action

LITERACY LINK

READING

After reading this lesson, write five questions about carbon capture that you would like to have answered.

NUMERACY LINK

CALCULATION

For every 1000 kg of cement that is produced, 900 kg of CO_2 is emitted into the environment. Express this as a percentage.

How much CO_2 would be emitted from the production of 5 tonnes of cement?

There are many different fields of scientific research and technological development. All around the globe, scientists and engineers are creating new materials, technologies and inventions, many of which have the promise of improving the world in some way.

However, just because a new technology is invented, or a new material is developed, it doesn't mean we should use it right away. Scientists have to consider many different factors during their research. Let's consider one example.

1 Carbon capture helps reduce carbon dioxide emissions

One field of materials science that has received a lot of attention in the 21st century is **carbon capture**. Carbon dioxide (CO_2) is a vital gas for plants, but human technology creates far more of it than the environment can process. Excess CO_2 pollutes the atmosphere and is a major contributor to climate change, so we need to find ways to reduce CO_2 levels around the world.

Carbon capture is a common way of dealing with CO_2. It is an engineering process that compresses CO_2 into a liquid form, which can then be stored underground.

How might carbon capture be useful?

Figure 4.16 The major sources of carbon dioxide emissions are burning fossil fuels for electricity, heat and transport.

2 There are possible alternatives to carbon capture

While carbon capture helps reduce the amount of CO_2 pollution in the atmosphere, it's not a perfect solution. It's an expensive process that uses a lot of energy, and there's the risk of leakage from the storage sites. Because of these issues, scientists are also researching other ways to deal with carbon dioxide emissions.

In 2019, scientists from RMIT in Melbourne developed a low-cost technique for converting atmospheric CO_2 into solid particles of carbon. This process uses liquid metal as a catalyst, and adds energy in the form of an electric current. The chemical reaction causes the CO_2 to condense into flakes of carbon (which are much like coal) at room temperature.

Another useful material could be Ferrock, an alternative to concrete created by American scientist David Stone. Concrete is used in buildings throughout the world, but the process of making it emits large amounts of CO_2. Ferrock contains steel dust and silica (ground glass), and the process of making it actually uses up CO_2 instead of creating it.

Why are alternatives needed for carbon capture?

3 Many factors influence decisions about scientific research

Scientists need to consider the **implications** of new technologies and materials being put into widespread use. There are many different types of considerations.

- Social considerations: what impact could this have on society and the way people or organisations behave?
- Ethical considerations: is it morally right to use this new material?
- Environmental considerations: what impact could this new technology have on the environment?

Consider the solid carbon made from atmospheric CO_2. One of the potential uses for reformed carbon is as fuel – but that means that the captured CO_2 would be released into the atmosphere again. What impact would this have on the environment – would it help or actually make things worse?

Considerations such as this mean that scientific research must be done carefully. Scientists, organisations and governments need to think through the implications of every new technology.

Name three types of considerations that influence scientific research.

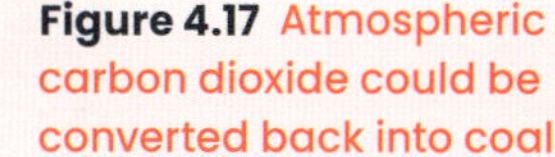

Figure 4.17 Atmospheric carbon dioxide could be converted back into coal.

CHECKPOINT 4.7

1 Why do chemists have an important role to play in the manufacture of new materials?

2 What do you think the statement 'scientists turn carbon dioxide into coal' means?

3 Why is it necessary to reduce the amount of CO_2 in the atmosphere?

4 Explain why Ferrock might have less environmental impact than concrete.

5 Describe three important factors to consider when conducting scientific research.

6 If scientists do turn carbon dioxide back into coal on a large scale, how could this help the environment? How could it hurt it?

ETHICAL CAPABILITY

7 Research the uses of fuel cells. What are the social, ethical and environmental implications of using fuel cells instead of combustion engines?

SUCCESS CRITERIA

- I can discuss how social, ethical and environmental considerations influence decisions about scientific research.

VISUAL SUMMARY

All matter is made up of atoms that have mass.

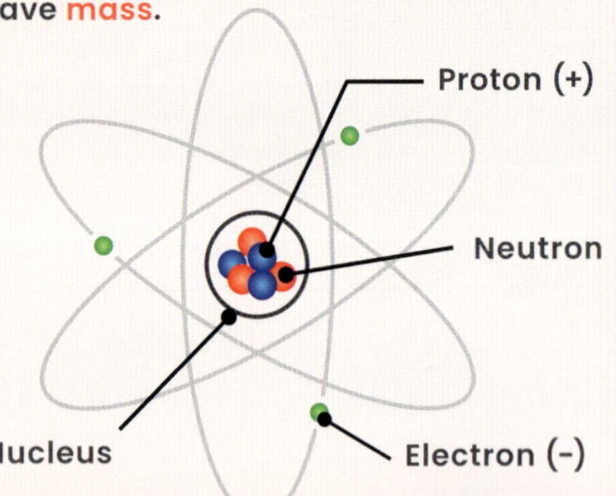

A **chemical equation** shows what atoms are involved before the reaction and how they recombine afterwards.

Chemical formulas:
- are a shorthand way of writing the name of an element or compound
- tell you the number and type of atoms in a compound
- can be used to write equations that model reactions.

Chemical reactions involve elements, compounds and energy.

Factors affecting reaction rates include:
- concentration
- temperature
- pressure
- surface area
- catalysts.

Atoms bond to form elements, covalent compounds or ionic compounds.

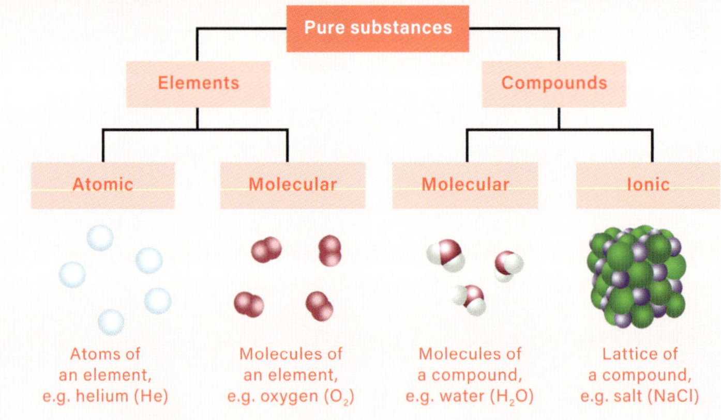

Increasing surface area increases reaction rate.

A large cube
- Smaller surface area
- Lower reaction rate

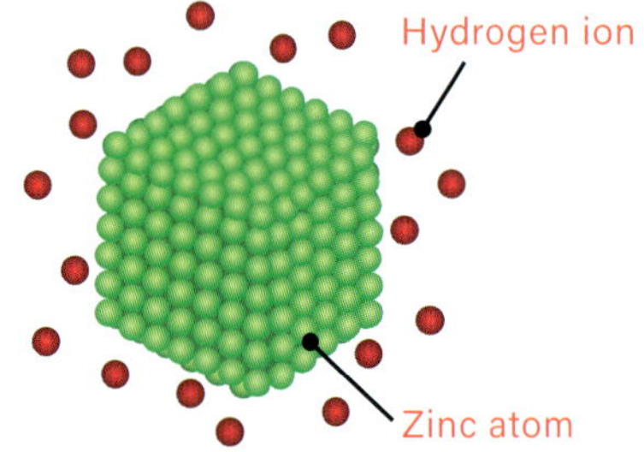

Smaller pieces
- Larger surface area
- Higher reaction rate

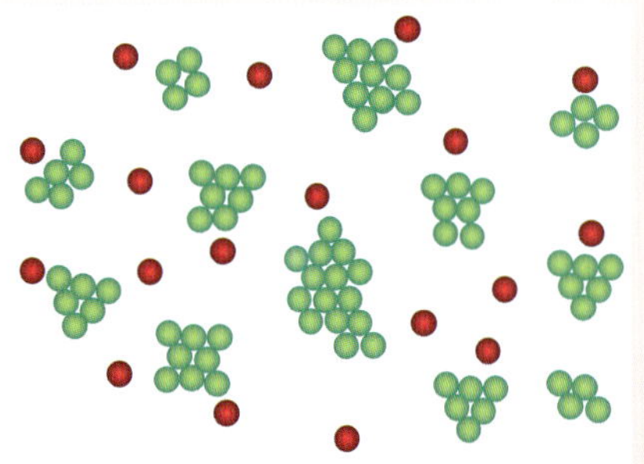

Scientists have to consider many different factors in their research.

- **Social considerations:** what impact could this have on society?
- **Environmental considerations:** what impact could this have on the environment?
- **Ethical considerations:** is it morally right?

★ FINAL CHALLENGE ★

1. Explain the difference between an element and a compound.
2. Draw a diagram of an atom and label the protons, neutrons and electrons.
3. Explain what an isotope is.

Level 1

50xp

LEVEL UP!

4. What do ionic and covalent compounds have in common, and how do they differ?
5. Balance the following chemical equations.
   - a $Fe + O_2 \rightarrow Fe_3O_4$
   - b $Mg + HCl \rightarrow MgCl_2 + H_2$
   - c $C_6H_{12}O_6 + O_2 \rightarrow CO_2 + H_2O$
6. Name some factors that increase the rate of reaction.

7. In your own words, summarise the law of conservation of mass and explain why this law is important in chemistry.
8. Explain what activation energy is and why it must be supplied to start almost all reactions.
9. Explain how compounds are named.

10. Explain, in terms of particles, how increasing the concentration of a reactant increases the rate of reaction.
11. Increasing the surface area of a reactant increases the rate of reaction. Explain why.
12. Describe the role of catalysts in increasing the rate of a chemical reaction.

13. Write the chemical formulas for copper nitrate, aluminium oxide and lead phosphate.
14. Discuss some of the social, ethical and environmental considerations in the manufacturing of plastic.

5 GLOBAL SYSTEMS

Why is it important that we recognise patterns within our world and beyond?

No matter how big or small, everything that happens on Earth has an effect on everything else. You can consider Earth to have four interacting spheres – the atmosphere (air), hydrosphere (water), lithosphere (rock) and biosphere (living things) – which help explain how the natural systems and cycles work. Knowledge of Earth's four spheres is especially important for understanding how human activity affects Earth and how we can preserve natural environments for future generations.

1 FRAYER MODEL

Copy and complete the below chart in your workbook.

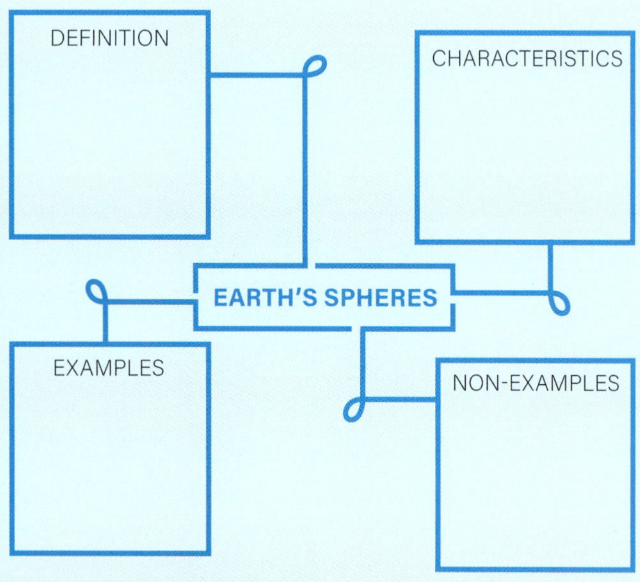

Complete two additional charts for the key terms *Climate change* and *Fossil fuels*.

2 LEARNING LINKS

Brainstorm everything you already know about global systems.

3 SEE-KNOW-WONDER

List three things you can SEE, three things you KNOW and three things you WONDER about this image.

4 CRITICAL + CREATIVE THINKING

ALPHABET KEY: Create an A–Z list of words associated with Earth's four spheres.

THE REVERSE: What would happen if carbon dioxide was a product of photosynthesis, not a reactant?

WHAT IF: All the ice on Earth melted overnight?

5 AN INSPIRING CLIMATE ACTIVIST!

Greta Thunberg is a young Swedish environmental activist who took the world by storm in 2018. Aged 15, Greta began protesting every Friday outside Swedish parliament, demanding action on climate change and inspiring millions of young people around the world to get involved in school strikes. Greta went on to speak at the United Nations asking people to 'act as though your house is on fire, because it is'. Greta has won countless awards for her activism, including two nominations for the Nobel Peace Prize. Go Greta!

5.1 EARTH'S SPHERES

LEARNING INTENTION

At the end of this lesson I will be able to describe Earth's atmosphere, hydrosphere, lithosphere and biosphere.

KEY TERMS

atmosphere
the layer of gas that surrounds Earth

biosphere
all of the living things on Earth

hydrosphere
all the water on Earth

lithosphere
Earth's rigid outer zone (crust and upper mantle), made up of tectonic plates

LITERACY LINK

VOCABULARY

Etymology is the study of the origin of words. The word *sphere* comes from the Greek word *sphaira* meaning 'ball'. Propose meanings for the words *hydro*, *atmos*, *lithos* and *bio*.

NUMERACY LINK

GRAPHING

Create a pie chart to show the allocation of fresh water on Earth.

- Glaciers and ice caps: 69%
- Groundwater: 30%
- Available fresh water: 1%

Earth scientists consider Earth to be a system made up of four interacting spheres – the atmosphere (air), hydrosphere (water), lithosphere (rock) and biosphere (life). By studying the spheres and how they interact, we can gain a better understanding of how Earth works and how humans are affecting it.

1 The atmosphere is a layer of gas

The **atmosphere** is the layer of gas that surrounds Earth. The atmosphere is made up of four major layers.

- The troposphere is the bottom layer and is where we live, jet planes fly and weather takes place.
- The next layer is the stratosphere, which contains the ozone layer.
- Above the stratosphere is the mesosphere, where meteorites burn up.
- The upper layer is the thermosphere, which is where aurora happen.

The atmosphere contains oxygen, which we breathe, and carbon dioxide, which plants need for photosynthesis. Oxygen makes up about 21% of the gases in the atmosphere. Other gases are nitrogen (78%), argon (0.93%) and carbon dioxide (0.04%). The atmosphere also contains minute quantities of neon, helium, methane, water vapour, krypton, hydrogen, xenon and ozone.

The atmosphere protects life on Earth from harmful radiation by either absorbing it or reflecting it back into space.

What is the atmosphere?

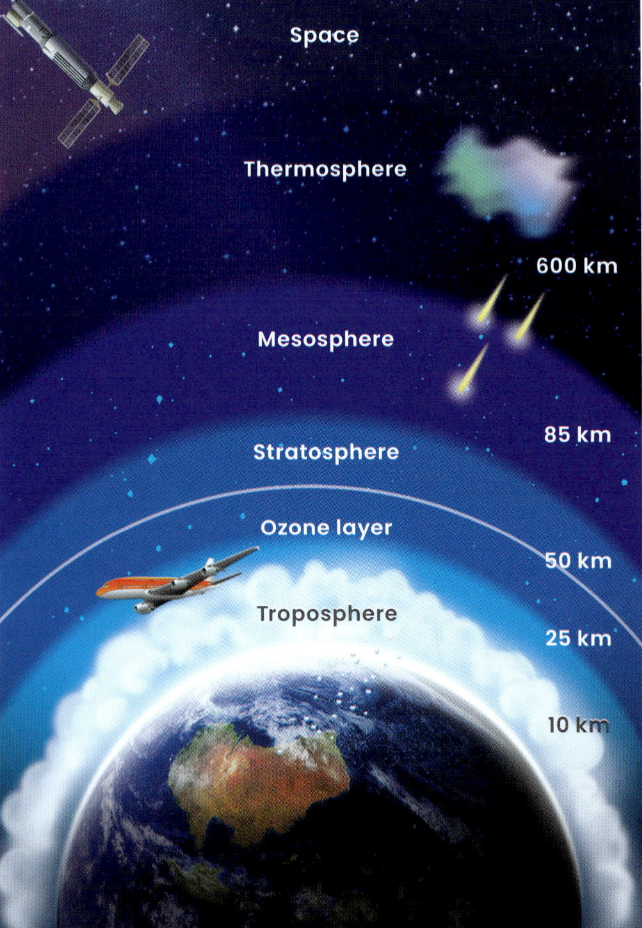

Figure 5.1 Earth's atmosphere is made up of four major layers: the troposphere, stratosphere, mesosphere and thermosphere.

❷ The hydrosphere is all the water on Earth

The **hydrosphere** is all the water on Earth, in all its forms – solid (snow/ice), liquid and gas (water vapour). The water is located on the surface (in oceans, rivers and lakes), in the atmosphere (clouds and water vapour), underground (groundwater) and in the bodies of living things (biological water).

More than 97% of the water on Earth is saline (salty). Less than 3% of the water on Earth is fresh water. Most of this fresh water is in glaciers and ice caps (69%), or as groundwater (30%). The remaining 1% is found in ground ice and permafrost, lakes, the atmosphere, living things, rivers, swamps, marshes and soil. All this water is connected through the water cycle.

What is the hydrosphere?

❸ The lithosphere is Earth's crust and upper mantle

The **lithosphere** is made up of Earth's crust and the top 100 km of the rigid upper mantle. It includes the rocks and soil, and landscape features such as mountains and valleys. The lithosphere is made up of 8 major and 7 minor tectonic plates that 'float' on the upper mantle. Movement of these plates results in volcanic eruptions and earthquakes.

What is the lithosphere?

❹ The biosphere includes all living things on Earth

The **biosphere** includes all the living things on Earth, from microbes to humans. When studying the biosphere, scientists consider how populations of different species interact with each other and their environment.

Biologists use food chains and food webs to represent interactions between organisms in ecosystems. Taxonomists use different levels to classify living things to show how they are related.

How do plants (in the biosphere) interact with the other spheres?

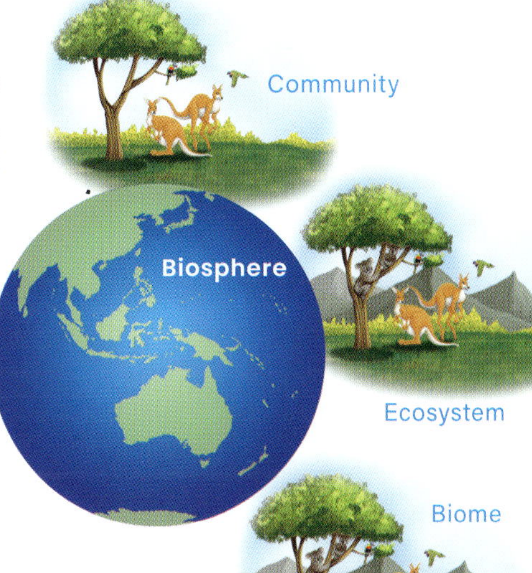

Figure 5.2 The biosphere includes all living things on Earth, how they interact with each other and their environment.

INVESTIGATION 5.1
Observing interactions between the spheres

KEY SKILL
Writing a research question

▶ *Go to page **156***

CHECKPOINT 5.1

1 How does the atmosphere protect Earth?

2 Identify the layer of the atmosphere where:
a we live
b the ozone layer is
c meteorites burn up
d aurora occur.

3 Describe the different ways in which water is stored in the biosphere.

4 Identify features of Earth that are part of the lithosphere.

5 Identify the sphere(s) that are involved in:
a a thunderstorm
b a wombat digging a burrow
c humans burning fossil fuel
d waves eroding a beach.

6 Suggest at least three ways in which the biosphere and hydrosphere interact.

INQUIRY

7 Brainstorm ways that cyclones impact on the four spheres: the atmosphere, the hydrosphere, the lithosphere and the biosphere.

SUCCESS CRITERIA

- I can describe the atmosphere, hydrosphere, lithosphere and biosphere.
- I can discuss some ways in which the four spheres interact.

5.2 THE CARBON CYCLE

LEARNING INTENTION

At the end of this lesson I will be able to explain how the carbon cycle is an example of the connection between Earth's spheres.

KEY TERMS

carbon cycle
the cycle that explains how carbon moves between Earth's spheres

cellular respiration
the process that all living things use to produce cellular energy from glucose and oxygen

fossil fuel
a fuel that is formed from the decomposition of dead animals and plants over millions of years; e.g. oil and coal

photosynthesis
the process that plants use to make glucose from carbon dioxide and water

LITERACY LINK

WRITING

Rewrite section **1** with the aim of making the language more formal and scientific. Create a new paragraph that does not use everyday or informal language and uses scientific terminology wherever possible.

Matter cannot be created or destroyed – instead, it is recycled. One day, an atom is part of the atmosphere; the next day, it's part of a plant. A week later, that atom is part of an animal, and eventually it moves into the soil and then the ocean. The same atoms are recycled over and over again as they move through the four spheres.

We use models called cycles to better understand the movement of matter between the spheres. The carbon cycle is a particularly important cycle that affects all of Earth's systems.

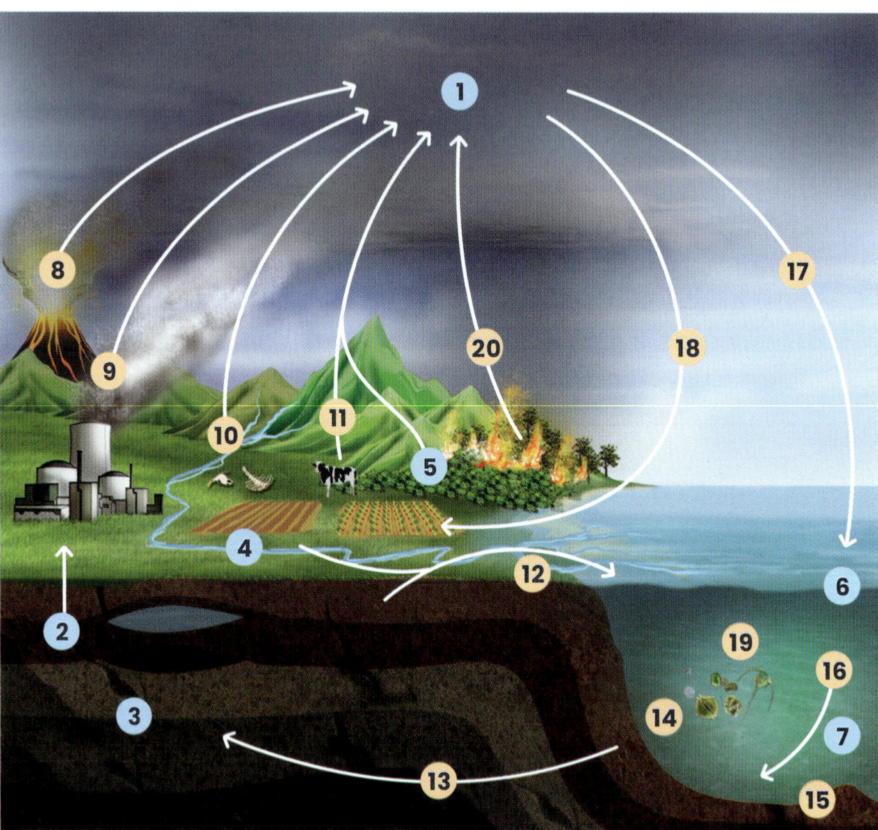

CARBON STORES
1 Atmosphere
2 Coal, oil, gas
3 Sediments and sedimentary rock
4 Soil and organic matter
5 Plants
6 Ocean surface
7 Deep ocean

PROCESSES	
8 Erupting volcanoes	**15** Sinking sediment
9 Burning fossil fuels	**16** Deep ocean currents
10 Decomposition	**17** Carbon dioxide exchange
11 Respiration	**18** Photosynthesis
12 Weathering, erosion and run-off	**19** Phytoplankton
13 Rock formation	**20** Burning
14 Shellfish and corals	

Figure 5.3 The carbon cycle shows how atoms of carbon move between the atmosphere, hydrosphere, lithosphere and biosphere.

❶ Carbon is a vital element

Carbon is a key element for life. It is the second most abundant element (after oxygen) in the human body and is just as important for other living things. Carbon is the main building block of carbohydrates, fats and proteins – which make up pretty much everything in your body. When you eat food, you are consuming carbon. When you exhale, you are removing carbon from your body in the form of carbon dioxide.

The **carbon cycle** explains how carbon moves between Earth's spheres. In the atmosphere, carbon is found in molecules such as carbon dioxide, which helps to regulate Earth's temperature. In the lithosphere, carbon is found in soils as decomposing matter, and in rocks, often in the form of calcium carbonate ($CaCO_3$). In the hydrosphere, carbon is in the form of carbon dioxide dissolved in oceans, lakes and rivers.

Why is carbon a key element for life?

INVESTIGATION 5.2A

Releasing dinosaur breath

KEY SKILL

Explaining results using scientific knowledge

▶ *Go to page **157***

INVESTIGATION 5.2B

The effect of temperature on soil respiration

KEY SKILL

Identifying the independent, dependent and controlled variables

▶ *Go to page **158***

❷ Carbon cycles through Earth's spheres

Carbon moves between Earth's spheres through some of these processes:

- Photosynthesis moves carbon from the atmosphere into plants and other producer organisms. **Photosynthesis** is the process by which plants synthesise glucose, which can then be used for growth and respiration.
- Carbon moves through food webs from producers to consumers.
- **Cellular respiration** moves carbon from the biosphere to the atmosphere. In cellular respiration, living things use oxygen and glucose to produce energy, with carbon dioxide being a waste product.
- Decomposition moves carbon from living things into the soil. When living things die and start to decay, their remains become part of the soil. Decomposers in the soil consume the dead matter and release the carbon back into the atmosphere through cellular respiration. If the dead matter is not decomposed, it can form **fossil fuels** such as oil and coal, locking the carbon away for hundreds of millions of years.
- Carbon dioxide dissolves into bodies of water where the surface mixes with the atmosphere. Here it often forms carbonate ions (CO_3^{2-}).
- Marine organisms use the carbon dissolved in sea water (in the form of carbonate ions) to build their shells.
- Tectonic movements bring rocks containing carbon, such as limestone (calcium carbonate) or fossil fuels (e.g. coal), to the surface. The carbon can then be washed into waterways, or returned to the atmosphere.
- Burning of forests and fossil fuels releases carbon into the atmosphere.
- Volcanic eruptions release carbon from the lithosphere into the atmosphere as carbon dioxide.

How can carbon move between the biosphere and the hydrosphere?

Figure 5.4 Decomposition releases carbon back into the atmosphere through cellular respiration.

5.2 continued ...

... 5.2 continued

THE CARBON CYCLE

KEY TERMS

carbon sink
a place where carbon is stored in the carbon cycle

greenhouse effect
the trapping of the Sun's warmth by the atmosphere

greenhouse gas
a gas that traps heat energy in the atmosphere

permafrost
frozen soil in the Arctic

NUMERACY LINK

DATA

Use the graph to answer the following questions.

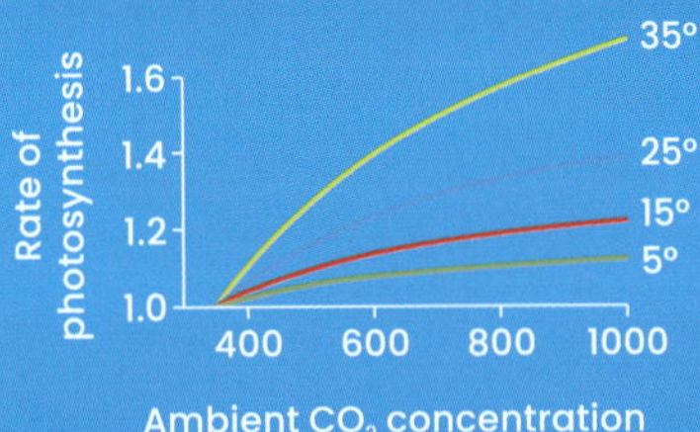

a Describe the pattern of results for this graph.

b Construct a table showing the rate of photosynthesis at different temperatures when the ambient CO_2 concentration is 600 μmol mol^{-1}.

❸ Carbon sinks are stores of carbon

Carbon sinks are where carbon is taken from the atmosphere and stored within the carbon cycle. This means that the carbon is locked up and cannot cycle for long periods of time. Examples of carbon sinks are forests and the ocean, although fossil fuel deposits and rocks that contain large amounts of calcium carbonate, such as limestone, are also significant.

Soil is one of the most important carbon sinks because more carbon is stored in soil than in living things and the atmosphere combined. This is because a lot of decomposed organisms end up as part of the soil. Once a carbon atom is locked up in the lithosphere, it can be buried for many millions of years until natural movement of the tectonic plates brings it up to the surface again.

The ocean is the world's largest carbon sink. When air from the atmosphere mixes with water in the ocean, carbon dioxide gas dissolves into the water. This carbon can then be used by marine creatures, such as molluscs and corals, to build their shells and skeletons.

What is a carbon sink?

Figure 5.5 Forests and oceans are places where carbon can be stored for long periods of time. These are known as carbon sinks.

❹ Carbon sources release carbon

Carbon sources release carbon into the atmosphere as part of the carbon cycle. Volcanoes, decomposition, respiration and fires are all natural carbon sources. Weathering of rocks containing calcium carbonate can also release carbon back into the atmosphere.

Decomposers, such as bacteria and fungi, play an important role in the carbon cycle. They return the carbon locked up in the bodies of dead organisms to the atmosphere as carbon dioxide, by consuming them and undergoing cellular respiration.

What is a carbon source?

Figure 5.6 Coal-fired power has significantly contributed to the increased amount of carbon in the atmosphere.

5 Humans influence the carbon cycle

The processes of mining and burning fossil fuels are interrupting the natural carbon cycle. Instead of the carbon atoms being locked away in sinks, they have been brought to the surface and added back into the active carbon cycle much faster than they would naturally.

This activity has greatly increased the amount of carbon in the atmosphere, mostly as carbon dioxide, but also as other **greenhouse gases** such as methane (CH_4), resulting in the enhanced **greenhouse effect**. Natural processes of the carbon cycle that remove carbon from the atmosphere and store them in sinks, such as photosynthesis, are unable to significantly reduce this excess carbon. Deforestation has added to this problem by reducing the size of forests and limiting the amount of photosynthesis that can take place.

Scientists have found that the ocean has absorbed some of the excess carbon from the atmosphere. However, this has led to ocean water becoming more acidic, having negative impacts on marine life. If ocean temperatures increase, the ability for the oceans to act as a carbon sink will decrease, as they will be unable to absorb as much carbon dioxide from the atmosphere.

There is a real threat that warming caused by the enhanced greenhouse effect will cause **permafrost** to thaw, releasing even more carbon into the atmosphere as the trapped methane and carbon dioxide it holds escapes.

How does the burning of fossil fuels change the natural carbon cycle?

CHECKPOINT 5.2

1 Identify the natural processes illustrated in Figure 5.3.

2 Identify the human-influenced processes illustrated in Figure 5.3.

3 What is the role of the carbon cycle?

4 Describe one way that carbon is moved from one sphere into another.

5 Create a flow chart to illustrate the effects of burning fossil fuels on the natural carbon cycle.

6 Another name for a matter cycle like the carbon cycle is a biogeochemical cycle. Propose why this is a suitable name.

7 Deforestation also affects the natural carbon cycle. Use information provided in this section to explain how.

EXTENSION

8 Explain how the oxygen cycle is linked with the carbon cycle.

SUCCESS CRITERIA

- I can describe four major steps in the carbon cycle.
- I can describe how the carbon cycle connects the atmosphere, hydrosphere, lithosphere and biosphere.

5.3 THE OZONE LAYER

LEARNING INTENTION

At the end of this lesson I will be able to describe how social actions have led to changed government policies and social-behavioural change in relation to the use of chlorofluorocarbons (CFCs) in aerosol spray cans.

KEY TERMS

ozone
a molecule made of three oxygen atoms bonded together

ozone-depleting substance
a chemical that reacts with ozone and breaks the molecule down as well as stopping the molecule from re-forming

ozone layer
a region of the stratosphere with a high concentration of ozone

ultraviolet radiation
harmful rays from the Sun, which can cause cancers and harm animals and plants

NUMERACY LINK

CALCULATION

In 2000, the ozone hole above the Antarctic was approximately 24.8 million km^2 and by 2017 it had decreased to 17.4 million km^2. Calculate the percentage change in the ozone hole.

The ozone layer is a part of the stratosphere that protects life from harmful ultraviolet (UV) radiation. In 1976, it was found that chemical pollution from industry reacted with the ozone in the ozone layer, reducing ozone levels and preventing more ozone from forming. This pollution resulted in a world-wide depletion of ozone in the stratosphere. The depletion is especially high over Antarctica, where a 'hole' has formed in the ozone layer. This reduction in ozone means that more UV radiation can reach the surface, harming life forms. Once the cause of the ozone hole was determined, countries acted quickly to ban the use of ozone-depleting chemicals, which means the ozone hole has begun to close over.

1 The ozone layer absorbs UV radiation

Ozone (O_3) is a molecule made of three oxygen atoms bonded together. Ozone is found throughout the atmosphere, but is especially concentrated in the **ozone layer**.

Ozone absorbs harmful **ultraviolet (UV) radiation**, and prevents it from getting to Earth's surface. There are different wavelengths of UV radiation, which have different effects. Ozone absorbs all of the short-wave UV-C (the most harmful), and 90% of the medium-wave UV-B (which causes sunburn), but lets through 50% of the least harmful long-wave UV-A. Without ozone in the atmosphere, life on Earth could not survive.

In what layer of the atmosphere is the ozone layer found?

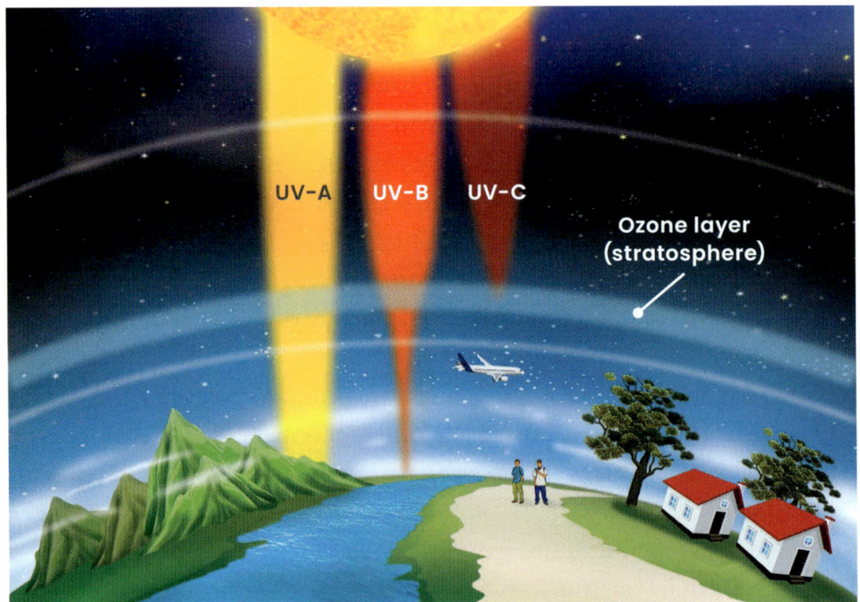

Figure 5.7 The ozone layer is a layer in the stratosphere where ozone (O_3) is abundant. The ozone layer prevents most of the harmful UV radiation from reaching Earth's surface.

2 Ozone is depleted by substances in pollution

In 1976, atmospheric researchers found that the ozone layer was becoming depleted by substances released by industry, such as chlorofluorocarbons (CFCs). CFCs were once commonly used in air conditioners and as a propellant for aerosol sprays, such as hairspray and deodorant. When CFCs are released into the atmosphere, they make their way into the stratosphere. Here they react with ozone, breaking some bonds and releasing O_2. They also prevent new ozone from forming.

Chemicals that react in this way are known as **ozone-depleting substances**. Most ozone-depleting substances stay in the stratosphere for decades. The ozone layer has reduced all over Earth by about 5%, which allows more UV radiation through.

What are the major chemicals that deplete the ozone layer?

3 The Montreal Protocol was designed to protect the ozone layer

Satellites and weather balloons have collected data about the amount of ozone in the stratosphere since the 1970s. In 1985, scientists published results about a hole in the ozone layer over the South Pole (Antarctica). People all around the world were shocked into action and determined to change policies that allowed CFCs in aerosol sprays.

In 1987, as part of the Montreal Protocol, all countries agreed to reduce and phase out their use of ozone-depleting substances. Different countries had different emissions targets depending on their economies and their current use of ozone-depleting substances. CFCs were relatively easy to replace, and most countries banned the use of CFCs in the 1990s. However, many industries replaced CFCs with hydrofluorocarbons (HFCs), which are potent greenhouse gases. The Protocol was amended to include phasing out HFCs.

The Montreal Protocol is one of the most successful international environmental agreements. The amount of ozone-depleting chemicals in the atmosphere has decreased, and the ozone layer has shown some signs of recovery. The hole is expected to close by 2050.

What does the Montreal Protocol require countries to do?

Figure 5.8 Data collected via weather balloons and satellites inform scientists about the health of the ozone layer.

INVESTIGATION 5.3
Investigating sunscreens

KEY SKILL
Evaluating results for reliability and validity

▶ *Go to page **160***

CHECKPOINT 5.3

1 What is the chemical formula for ozone?
2 Why is the ozone layer important?
3 Describe what an ozone-depleting substance does to the ozone layer.
4 Identify the source of the increased levels of ozone-depleting substances in the atmosphere.
5 How did social action lead to change in CFCs in aerosol sprays?
6 What is the relationship between the Montreal Protocol and the amount of ozone-depleting substances in the atmosphere?

CONNECTING IDEAS

7 Because of the circulation of the atmosphere and the location of the continents, an Arctic (North Pole) ozone hole wasn't observed until 2010–2011. Do you think a Northern Hemisphere ozone hole has different consequences from the one over Antarctica? Justify your response.

SUCCESS CRITERIA

- I can describe the importance of the ozone layer.
- I can identify how human activity has depleted the ozone layer.
- I can outline how social action led to change in CFCs in aerosol sprays.

5.4 THE ENHANCED GREENHOUSE EFFECT

LEARNING INTENTION

At the end of this lesson I will be able to evaluate scientific evidence for the effect that human activity has on the greenhouse effect.

KEY TERMS

enhanced greenhouse effect
an increase in the greenhouse effect due to human greenhouse gas emissions

greenhouse effect
the trapping of the Sun's warmth by the atmosphere

greenhouse gas
a gas that traps heat energy in the atmosphere

LITERACY LINK

SPEAKING

Identify an issue or argument that is preventing action on reducing greenhouse gas emissions. Research and prepare some dot points that could be used to counter this issue or argument. Present your rebuttal to your class.

A greenhouse is a glass structure that gardeners use to trap the Sun's energy and keep their plants warm. The glass allows the energy through but prevents some of it from escaping. Earth's atmosphere acts in a similar way. Carbon dioxide is a significant greenhouse gas, and emissions from the burning of fossil fuels have increased the amount of carbon dioxide in the atmosphere, artificially enhancing the greenhouse effect and raising the average temperature of Earth.

1 The greenhouse effect keeps Earth's surface warm

Without the **greenhouse effect**, the surface of Earth would be very cold. When the Sun's radiation hits the surface of Earth, some of it is reflected back out into space and some is absorbed by Earth's surface, warming it. The warm surface emits energy as infrared radiation.

A **greenhouse gas** is a gas that stops this infrared energy from going straight back out into space. Molecules of greenhouse gases absorb and then re-emit this energy, warming the lower atmosphere and the surface of Earth. The major greenhouse gases in the atmosphere are water vapour (H_2O), carbon dioxide (CO_2), methane (CH_4), nitrous oxide (N_2O) and ozone (O_3).

Why is the greenhouse effect important for life on Earth?

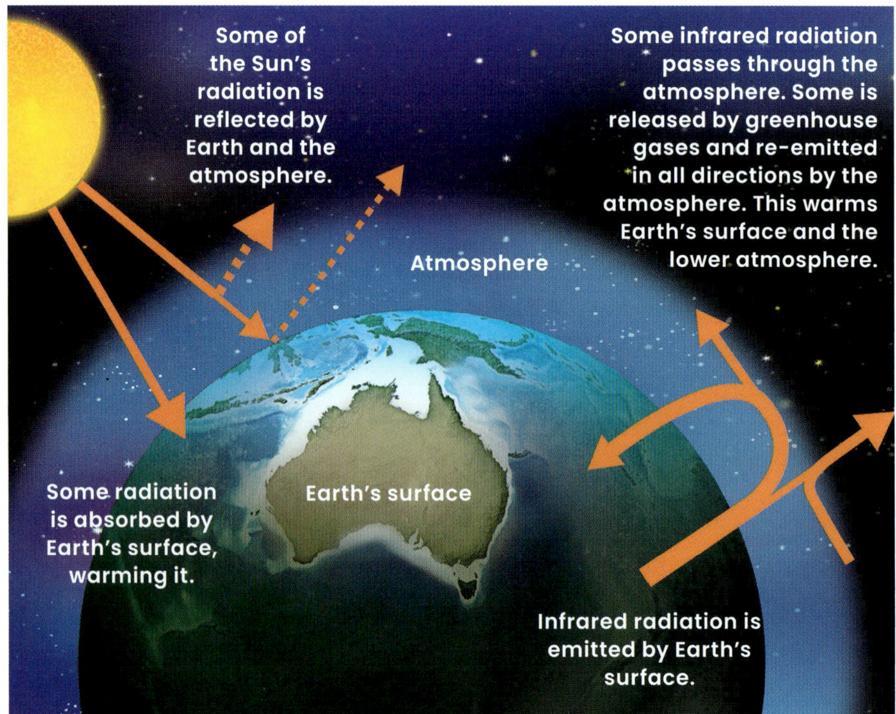

Figure 5.9 The greenhouse effect is important for maintaining the temperature on the surface of Earth, but increasing levels of greenhouse gases due to human activity have increased Earth's surface temperature.

❷ Humans have increased the greenhouse effect

Human activity, such as the burning of fossil fuels and factory emissions, increases the levels of greenhouse gases in the atmosphere. Because there are more gas particles in the atmosphere, this means that more infrared radiation stays in Earth's atmosphere and less is radiated back into space. This has increased Earth's surface temperature. This is called the **enhanced greenhouse effect**.

How have humans enhanced the greenhouse effect?

INVESTIGATION 5.4
The greenhouse effect

KEY SKILL
Representing data to identify patterns and trends

▶ *Go to page **162***

❸ Atmospheric measurements show CO_2 levels are increasing

Scientists can measure the amounts of the greenhouse gases in the atmosphere and correlate this with average temperature curves to demonstrate the relationship between the two. A global network of stations samples the atmosphere regularly. The most well-known station, Mauna Loa Observatory on the Big Island of Hawaii, has been operating since the 1950s.

The Cape Grim Baseline Air Pollution Station in north-west Tasmania was established in 1976. Its location has some of the cleanest air in the world, because it is well away from sources of pollution. Measurements taken at Cape Grim show that carbon dioxide levels in the atmosphere are increasing. In 2019, the atmospheric concentration of carbon dioxide was about 405 ppm.

Why is Cape Grim a good place to take air samples?

Figure 5.10 The Cape Grim Baseline Air Pollution Station samples the air coming off the Southern Ocean.

5.4 continued ...

❹ The Keeling curve plots CO_2 levels in the atmosphere

In 1958, US scientist Charles Keeling started monitoring carbon dioxide concentrations in the atmosphere at Mauna Loa, Hawaii. He was the first person to take regular measurements that showed how the concentration of carbon dioxide in the atmosphere was changing.

Keeling's measurements showed a steady increase in carbon dioxide concentration in the atmosphere and that the amount of carbon dioxide in the atmosphere changes throughout the year.

The annual cycle is because most of Earth's land mass and plant life are in the Northern Hemisphere. Over the northern spring and summer, plant growth and photosynthesis increase, which reduces carbon dioxide in the atmosphere. During autumn and winter, plant growth and photosynthesis decline and levels of carbon dioxide in the atmosphere increase. When plotted as a graph, this data is known as the Keeling curve.

Where did Keeling gather his data on atmospheric carbon dioxide concentrations?

... 5.4 continued

THE ENHANCED GREENHOUSE EFFECT

NUMERACY LINK

DATA

CO_2 (parts per million)
400
390
380
370
360
350
340
330
320
310
1965 1975 1985 1995 2005 2015
Year

Annual cycle of CO_2
Jan April July Oct Jan

Use the graph to:

a determine what the concentration of atmospheric CO_2 concentration was in 1980

b determine what year the concentration of CO_2 reached 370 ppm

c predict the atmospheric CO_2 concentration in 2030.

❺ Ice core data also shows CO_2 levels are increasing

Ice cores are cylinders of ice drilled from ice sheets. Scientists use ice cores from Greenland (up to 123 000 years old) and Antarctica (up to 800 000 years old) to find out more about Earth's past temperature and climate. Importantly, they can analyse air bubbles trapped in the ice to determine the concentration of gases in Earth's atmosphere over time.

Data from the Vostok and Law Dome ice cores in east Antarctica shows that although carbon dioxide levels have varied in the past, the recent increase in carbon dioxide levels due to human activity is unprecedented.

What do scientists analyse from ice cores?

Figure 5.11 Combined data from the Vostok and Law Dome ice cores and Mauna Loa show how the carbon dioxide concentration in the atmosphere has varied over the last 400 000 years.

❻ Impacts of an enhanced greenhouse effect

As the level of carbon dioxide in the atmosphere has risen, the average temperature of Earth has increased. This has led to many changes to Earth's spheres, all of which have flow-on effects.

As the oceans get warmer, more water evaporates and atmospheric circulation patterns change. This means that cyclones and hurricanes can form further away from the equator, and that storms become larger as more water moves into the atmosphere. Rainfall patterns also change, so some areas receive more rain and other areas receive less.

Sea ice coverage in the Arctic has decreased. Solar radiation is reflected by light-coloured surfaces such as ice, and absorbed by darker surfaces. The reduction of ice has caused the Arctic Ocean to absorb more solar radiation, further warming the ocean and decreasing the sea ice. This, along with the melting of any land-based ice, can lead to sea levels rising significantly.

As carbon dioxide levels in the atmosphere increase, more carbon dioxide is absorbed by the ocean. This forms carbonic acid, making the oceans more acidic.

What is one way the enhanced greenhouse effect is impacting on Earth's systems?

Figure 5.12 The enhanced greenhouse effect describes the result of increased carbon dioxide levels in the atmosphere. Less heat is escaping into space so the temperature of Earth is rising, which has significant flow-on effects on aspects of our environment such as the climate and sea levels.

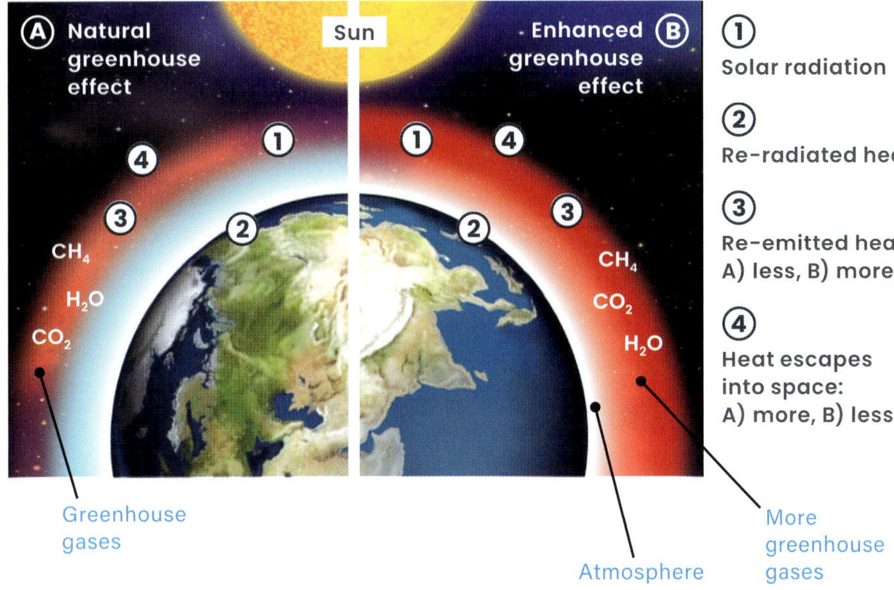

7 The Kyoto Protocol and Paris Agreement are attempts to reduce greenhouse gas emissions

The Intergovernmental Panel on Climate Change (IPCC) is a United Nations group that provides scientific information about climate change, its impacts and ways to respond. The IPCC has stated that reducing emissions of carbon dioxide and other greenhouse gases will reduce the concentration of these gases in the atmosphere, and thus reduce global temperature increases.

There have been several agreements by global leaders to address these issues and reduce the greenhouse gas emissions of their countries. The Kyoto Protocol (1997) and the Paris Agreement (2015) have been the major agreements between countries to reduce emissions to address climate change.

Barriers to reducing greenhouse gas emissions are primarily economical. For some countries, fossil fuels are still the cheapest form of energy. Many countries, including Australia, rely heavily on income from the mining and sale of fossil fuels. Across the world, some governments are encouraging people to choose renewable energy options by subsidising the costs involved. Other governments encourage people to purchase electric or hybrid cars by removing taxes involved with buying those cars. Australia has pledged to reduce its emissions by 26–28% of its 2005 levels by 2030.

What are some ways the enhanced greenhouse effect can be reduced?

CHECKPOINT 5.4

1 List some common greenhouse gases.

2 Explain the difference between the enhanced greenhouse effect and the natural greenhouse effect.

3 Use your knowledge of the carbon cycle to explain how mining and burning fossil fuels affects Earth's spheres.

4 The annual fluctuations in carbon dioxide level shown on the Keeling curve have been likened to 'Earth breathing'. Explain why.

5 Explain how data from ice cores is significant when considering how carbon dioxide emissions are affecting the temperature of Earth.

6 Identify some issues that are preventing countries from reducing their greenhouse gas emissions.

INQUIRY

7 Brainstorm with a partner the ways that climate change (the enhanced greenhouse effect) could result in a loss of biodiversity. Suggest some impacts the loss of biodiversity would have.

SUCCESS CRITERIA

- I can explain the difference between the natural and enhanced greenhouse effects.
- I can describe how human activity has enhanced the greenhouse effect.
- I can discuss some of the evidence for the enhanced greenhouse effect and explain how it contributes to scientific knowledge.
- I can describe some impacts of the enhanced greenhouse effect.

VISUAL SUMMARY

Earth's spheres relate to air, water, rock and life on Earth.

The atmosphere is the 600 km thick layer of gas that surrounds Earth.

The hydrosphere is all the water on Earth in all its forms.

The lithosphere is Earth's crust and upper mantle.

The biosphere is all of the living things on Earth.

▲ The carbon cycle shows how carbon atoms are recycled as they move through the different spheres.

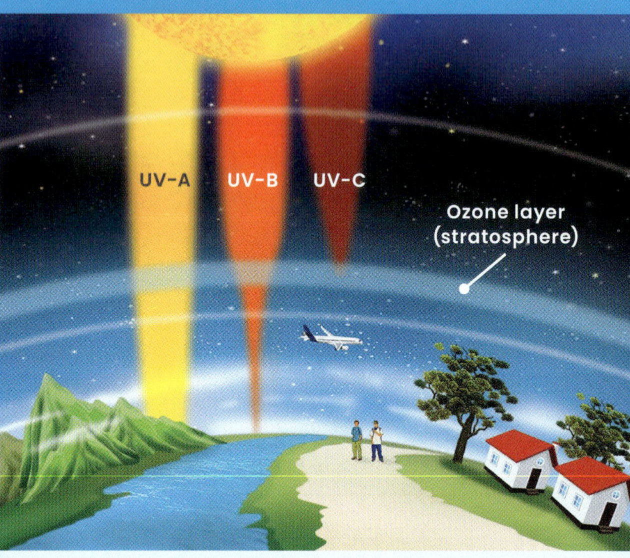

The ozone layer is a layer in the stratosphere where ozone (O_3) is abundant.

The ozone layer protects life from harmful UV radiation.

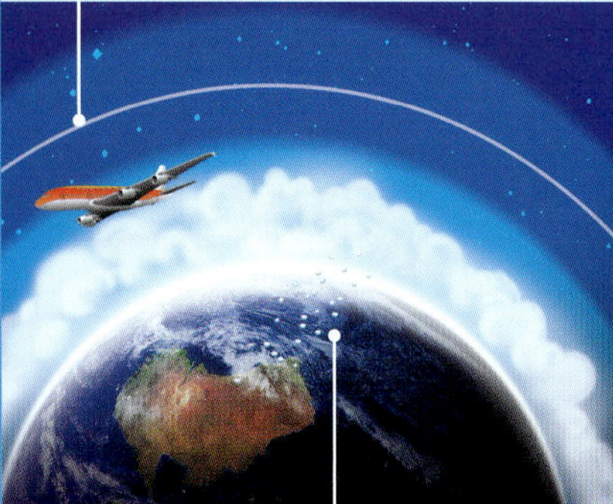

Carbon dioxide emissions are contributing to an enhanced greenhouse effect.

▼ Earth is kept warm by greenhouse gases that prevent heat being lost into space.

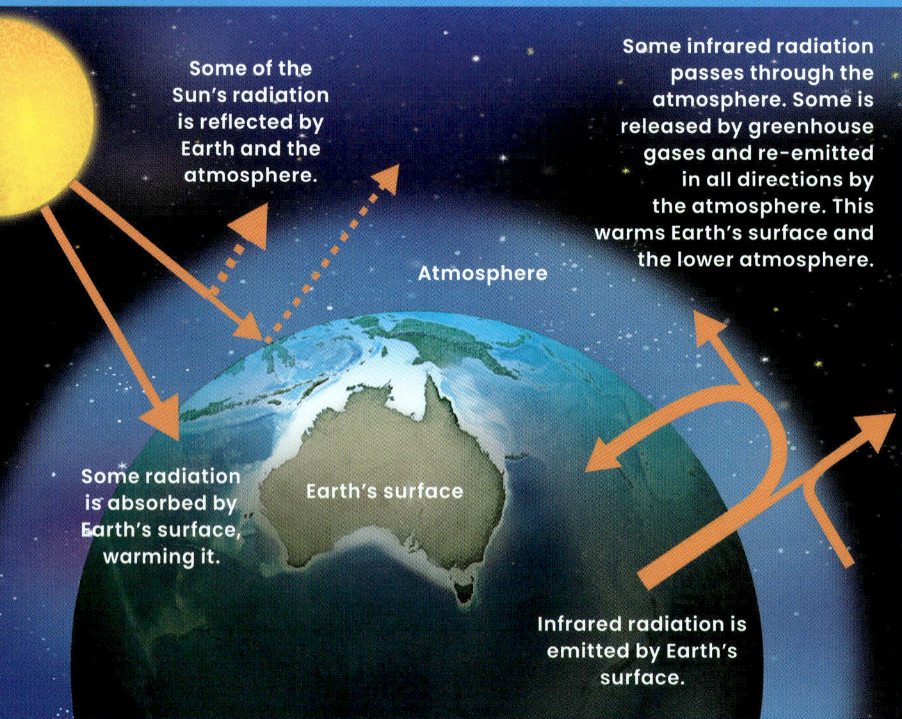

★ FINAL CHALLENGE ★

1. In one sentence each, describe the atmosphere, hydrosphere, lithosphere and biosphere.
2. Identify an example of an interaction that includes the:
   - a lithosphere and biosphere
   - b atmosphere and hydrosphere
   - c atmosphere and biosphere.

Level 1 ★☆☆ ☆☆☆ 50xp LEVEL UP!

3. Use a flow chart to show how carbon moves from one sphere to the next in the carbon cycle.
4. Explain the difference between the natural and enhanced greenhouse effects.
5. Explain what a carbon sink is, and give two examples.

Level 2 ★★☆ ☆☆☆ 100xp LEVEL UP!

6. Use an annotated diagram to illustrate how the greenhouse effect works.
7. Describe the relationship between the concentration of carbon dioxide in the atmosphere and the average surface temperature of Earth.
8. Explain why the Keeling curve is significant.
9. Provide these facts about the ozone layer:
   - a what it is composed of
   - b its effect
   - c how a hole was created in it
   - d what actions were taken to reduce the hole.

Level 3 ★★★ ☆☆☆ 150xp LEVEL UP!

10. Burning fossil fuels adds carbon dioxide to the atmosphere. Identify one impact of this on each of the atmosphere, the hydrosphere, the lithosphere and the biosphere.
11. Describe how human activity has led to the enhanced greenhouse effect.

Level 4 ★★★ ★☆☆ 200xp LEVEL UP!

12. Countries acted swiftly when it was discovered that CFCs and other chemicals were depleting the ozone layer. Although we know the impact that burning fossil fuels is having on the atmosphere, suggest why it has taken longer for countries to reach an agreement on how to address the issue.

Level 5 ★★★ ★★★ 300xp LEVEL UP!

THE UNIVERSE

Why is it important that we recognise patterns within our world and beyond?

Humans have gazed up at the night sky for millennia. Our ancestors observed that the sky changed with the seasons, used the stars to help them navigate, and linked their observations to stories of creation and mythology. We still look to the stars to help understand our place in the universe, to understand how it began and evolved, and how the Sun and solar system formed. As technology improves, scientists have been able to unlock and understand more of the mysterious universe.

1 FRAYER MODEL

Copy and complete the below chart in your workbook.

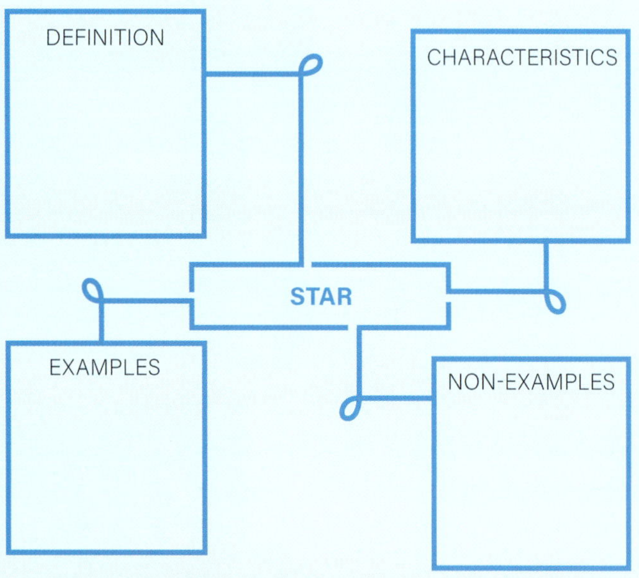

Complete two additional charts for the key terms *Big bang* and *Galaxy*.

2 LEARNING LINKS

Brainstorm everything you already know about the universe.

3 SEE-KNOW-WONDER

List three things you can SEE, three things you KNOW and three things you WONDER about this image.

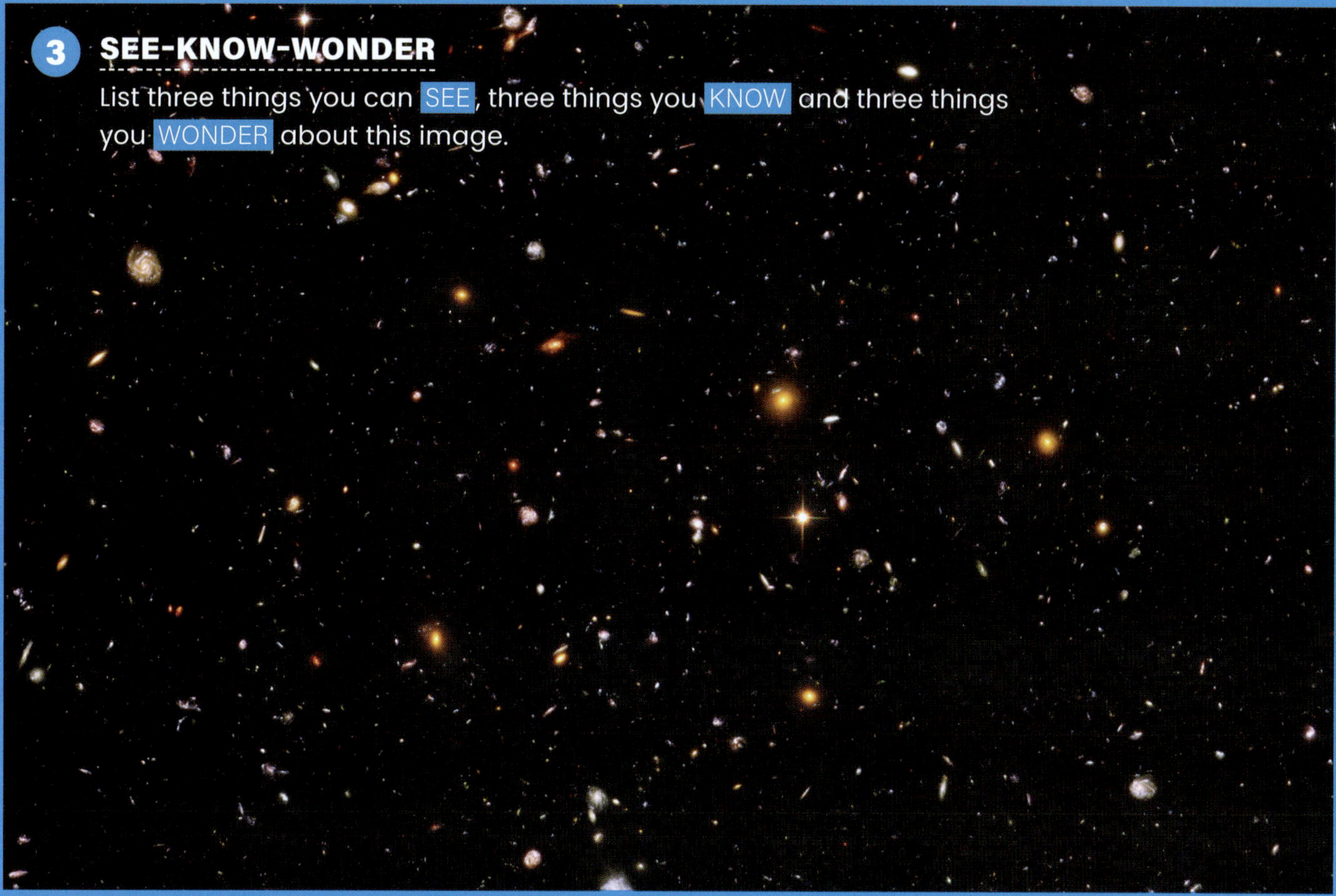

4 CRITICAL + CREATIVE THINKING

THE COMMONALITY: Outline some points of commonality between the big bang and a dinosaur.

VARIATIONS: List as many ways as possible that we could learn about the universe from Earth.

THE BRICK WALL: Think of ways to deal with the following situation: How to continue life on Earth without the Sun.

5 THE DENSEST MATTER!

You can imagine a black hole to be like a really heavy metal weight in the middle of a piece of fabric – the fabric is time and space and the metal weight is unimaginably dense matter left over from something like the death of a massive star. A black hole affects time and space, so things that are near a black hole start to slow down, lengthen and break apart. We are safe from black holes because there are none near Earth and they don't move around the universe gobbling things up like Pac-Man.

6.1 THE BIG BANG

LEARNING INTENTION

At the end of this lesson I will be able to use scientific evidence to outline how the big bang theory can be used to explain the origin of the universe and its age.

KEY TERMS

antimatter
particles that have properties opposite to that of normal matter

astrophysicist
a scientist who studies the physics of the universe

redshift
a change in light's wavelength towards the red end of the visible spectrum

singularity
an infinitely dense point of matter that existed before the big bang

LITERACY LINK

WRITING

Summarise the major steps in the history of the universe since the big bang, including the formation of the Sun and our solar system.

NUMERACY LINK

CALCULATION

The three most abundant elements in the universe are as follows: 74% hydrogen, 24% helium, 1% oxygen.

How many times more abundant is hydrogen than oxygen?

As technology has improved, and more observations and calculations have been made, so has our understanding of the universe and how it formed. The big bang theory explains how the universe rapidly expanded from a dense singularity about 13.7 billion years ago, creating matter and forming elements that then went on to create stars and galaxies.

1 The universe began as a singularity

The big bang theory is the currently accepted theory for the origin of the universe. The theory states that the universe began as a small dense region called a **singularity**, which then rapidly expanded about 13.7 billion years ago. This expansion is called 'the big bang'.

After the big bang, the first events happened very quickly. As it continued to expand, the universe started to cool, allowing the formation of particles of matter and **antimatter** (particles that are the opposite of matter). At about 0.001 seconds after the big bang, these particles annihilated each other. The matter that we observe today is what was left over from this interaction.

After 3 minutes, the protons and neutrons then combined to form the nuclei of atoms, about 75% hydrogen, 25% helium, and a small fraction of lithium. At this stage, all the matter and energy that would ever exist was formed, and all in the time it takes to boil a kettle. Between the formation of these nuclei and 500 000 years, the universe was a plasma of hydrogen, helium and lithium nuclei and free electrons.

At about 500 000 years, the universe cooled enough for the electrons to be attracted to the nuclei and so for atoms to form. This released energy in the form of photons of light. The first stars and galaxies then began to form 1 billion years after the big bang.

What does the big bang theory explain?

2 There are three pieces of evidence for the big bang

The universe is expanding

Edwin Hubble observed that light coming from distant galaxies was stretched into longer wavelengths – it was **redshifted**. This indicated that galaxies are moving away from us. The movements of the galaxies can be traced back to a single point, meaning that the universe must have once been contained in a small region of space.

Abundance of light elements

By mass, the universe is about 74% hydrogen and 24% helium, with the other 2% being all the other heavier elements. This abundance supports the big bang theory because if helium was only made by fusion in stars, there would be significantly less than 24%.

Cosmic microwave background radiation

Cosmic microwave background radiation is left-over heat released about 100 000 years after the big bang, when the universe had cooled enough to allow electromagnetic radiation to pass through it. Because the universe was not of uniform density when the cosmic microwave background was released, images show 'clumps' of matter. The cosmic microwave background would have originally been released as visible and UV light, but the expansion of the universe has redshifted it into the microwave band.

What is the evidence that the universe is expanding?

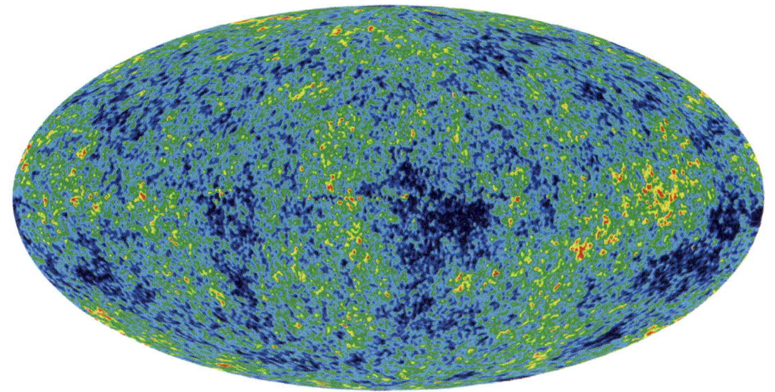

Figure 6.1 This image of cosmic microwave background radiation was taken from the Wilkinson Microwave Anisotropy Probe (WMAP).

3 Data from space telescopes helps scientists calculate the age of the universe

Current observations and calculations put the age of the universe at 13.7 billion years old. There are several pieces of evidence that have helped scientists to calculate this age.

The Hubble Space Telescope was able to measure how far away galaxies were from us and the speed that they were moving. This allowed **astrophysicists** to calculate how long it took galaxies to get to their current locations.

The Hubble Space Telescope was also used to calculate the ages of the oldest star clusters – the first stars that would have formed after the big bang. The data gathered by the Wilkinson Microwave Anisotropy Probe (Figure 6.2) also supports this age.

How can we tell how old the universe is?

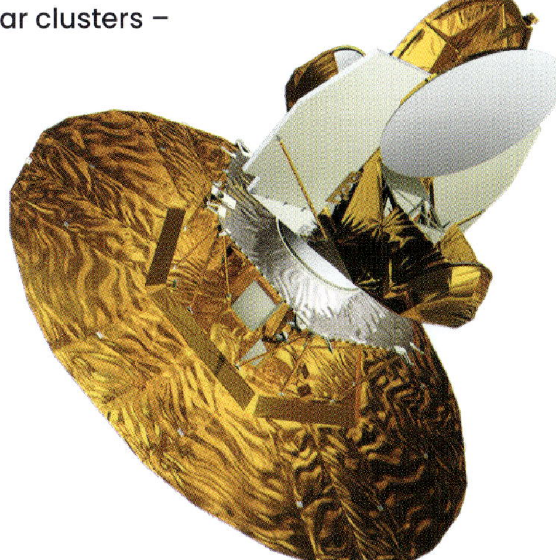

Figure 6.2 The Wilkinson Microwave Anisotropy Probe (WMAP) is a space-based microwave telescope.

INVESTIGATION 6.1

Modelling the expanding universe

KEY SKILL

Using modelling and simulations

▶ *Go to page **164***

CHECKPOINT 6.1

1 Describe the big bang theory.
2 What is a singularity?
3 When did atoms begin to form?
4 When did the first stars and galaxies begin to form?
5 Identify three pieces of evidence used to calculate the age of the universe.
6 Explain why the percentage of light elements in the universe supports the big bang theory.
7 Explain why the big bang is considered a theory.
8 What is the difference between matter and antimatter?
9 Explain what is meant by the following statement:
After three minutes, all the matter and energy that would ever exist was formed.

CONNECTING IDEAS

10 Create a timeline of major findings that helped us to understand how the universe formed.

SUCCESS CRITERIA

- I can describe the big bang theory.
- I can outline three major pieces of evidence that support the big bang theory.
- I can suggest evidence that was used to calculate the age of the universe.

6.2 STARS

LEARNING INTENTION

At the end of this lesson I will be able to outline some of the major features of stars and use appropriate scales to describe differences in size.

A star is a colossal mass of gas. Inside a star, elements are made through the process of nuclear fusion. Since the big bang, stars have come and gone in our universe. Small stars live for billions of years before expanding into **red giants**, then lose most of their mass to form a planetary nebula. Massive stars have shorter lives before expanding into red supergiants and ending as supernovas, which leave behind a neutron star or a black hole.

KEY TERMS

nebula
a vast region of gas and dust

neutron star
an extremely dense star left over after a supernova

red giant
a star that has stopped fusing hydrogen in its core

supernova
an explosion of a massive star at the end of its life

white dwarf
a small, very dense star formed at the end of a small star's lifetime

LITERACY LINK

WRITING

Write an interview with a white dwarf star about what they have experienced in their life.

NUMERACY LINK

UNITS

Convert the following numbers to scientific notation.

a 4200

b 0.0000065

1 Stars are mostly hydrogen

Hydrogen (H) is the most common element in the universe, and stars are mostly made of hydrogen. In the centre of stars, hydrogen atoms fuse together to form helium atoms. This reaction releases a lot of energy, including visible light.

This release of energy pushing outwards from the core counteracts the force of gravity, and so a star fusing hydrogen in its core is balanced. When these forces are not balanced, the star changes and moves through different stages of its life cycle.

What causes stars to make so much energy?

2 Nebulae are stellar nurseries

Nebulae are interstellar regions filled with gas and dust, and are where stars are born. A star begins to form when a region in the nebula starts to contract, and gravity brings the gas together. As more gas comes together, the mass starts spinning, forming a hot dense core called a protostar. The protostar continues to contract until the gas in the centre becomes dense enough and hot enough. This begins the fusion of hydrogen into helium, resulting in a star being born.

What is a nebula made of?

Figure 6.3 The Orion nebula is located near the tip of the sword in the constellation of Orion.

❸ The size of a star determines its life cycle

The size of a star determines how long it will live for, as well as the stages in its life cycle, because the size of the star affects how quickly all the hydrogen in its core fuses.

A star such as the Sun (1 solar mass) has a lifetime of about 10 billion years. Much larger stars (more than 7 solar masses) have much shorter life spans because they fuse hydrogen quickly. Smaller stars of less than 1 solar mass live a lot longer than the Sun because they fuse hydrogen more slowly.

A large star will have more gravity than a smaller star like the Sun. Anything in the universe that has mass has gravity – the more mass, the more gravity.

How long will the Sun live for?

❹ Stars eventually die

Once the amount of hydrogen within a star declines, the star enters its final phase and dies. This can happen in different ways depending on the star's mass.

A small star such as the Sun releases one last burst of energy, causing the outer layers of the star to puff outwards and form a planetary nebula. The star's core remains as an extremely hot, dense and small **white dwarf**.

In a large star, the atoms inside the core eventually fuse to become iron. More energy is required to produce iron than is given off, so gravity causes the star to rapidly contract and then explode. This is known as a **supernova**, and leaves the core of the star as an extremely dense **neutron star**.

In supermassive stars, the force of gravity is so strong that the core collapses. This forms a black hole, a body with such a strong gravitational field that it attracts all nearby matter, and even bends light around it.

How does a large star end its life?

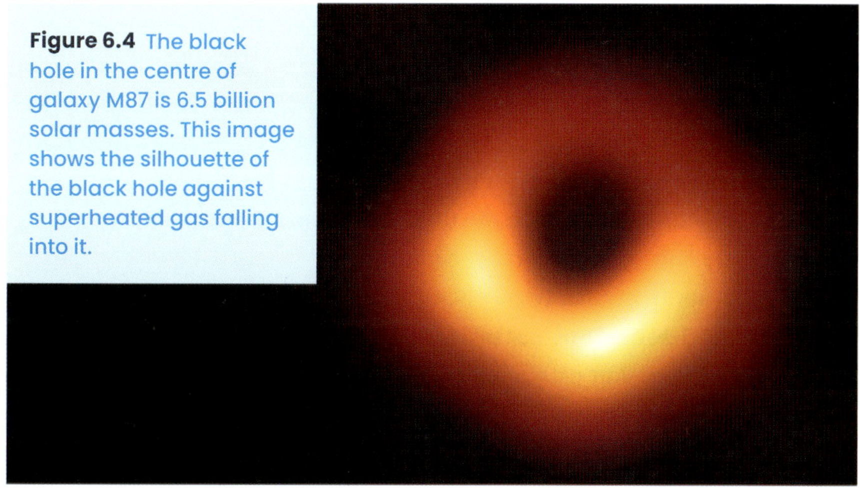

Figure 6.4 The black hole in the centre of galaxy M87 is 6.5 billion solar masses. This image shows the silhouette of the black hole against superheated gas falling into it.

INVESTIGATION 6.2
Stargazing

KEY SKILL
Referencing sources of information

▶ *Go to page **166***

CHECKPOINT 6.2

1 Identify the element that is produced in the core of the Sun.

2 Explain what might cause a nebula to start to contract.

3 What main factor determines the stages a star goes through?

4 Compare and contrast the lives of a sun-like star and a massive star.

5 Explain why solar mass is a useful measurement for comparing and contrasting stars.

6 'Planetary nebula' is the term used to describe what happens when the outer layers of a red giant are lost as it transforms into a white dwarf. Explain how this term is misleading.

7 Use diagrams to show how gravity acts on the matter inside a star at each point in its life cycle.

CONNECTING IDEAS

8 Use the information in this lesson, and further information from the internet, to explain the saying: 'You are made of stardust'.

SUCCESS CRITERIA

- I can describe how stars form and some of the key features of stars.
- I can use solar mass to discuss the difference in the sizes of stars.

6.3 GALAXIES

LEARNING INTENTION

At the end of this lesson I will be able to outline some of the major features of galaxies and use appropriate scales to describe sizes and distances between them.

Figure 6.5 The night sky over Kata Tjuta, looking towards the centre of the Milky Way galaxy.

KEY TERMS

astronomer
a scientist who studies space, stars and celestial objects

astronomical unit
the average distance between Earth and the Sun (about 150 000 000 km)

galaxy
a system of millions or billions of stars

light-year
the distance that light travels in one Earth year

parsec
3.26 light-years

LITERACY LINK

VOCABULARY

Identify some adjectives that can be used to describe galaxies.

NUMERACY LINK

CALCULATION

Calculate these distances in parsecs.

a Proxima Centauri to Earth: 4.3 light-years
b The centre of the Milky Way to Earth: 26 000 light-years
c Andromeda to Earth: 2.5 million light-years

Galaxies are massive groups of stars, gas, dust and other matter bound together by forces of gravity. The existence of galaxies beyond our own Milky Way was first proven by Edwin Hubble in the 1920s. Astronomers classify galaxies by their shape, and use the observations of many galaxies both close to and far away from our own Milky Way to discover how galaxies have formed, evolved and changed since the big bang.

1 The light-year is useful for measuring large distances

In the universe, distances are so great that the kilometre is too small to be a useful measurement. So, astronomers use the light-year to measure distances between stars and galaxies. A **light-year** (ly) is the distance light travels in one Earth year (365.25 days): 1 light-year is about 9.46 trillion kilometres (9.46×10^{12} km).

Our solar system is in the Milky Way galaxy and the closest galaxy to the Milky Way is Andromeda, which is 2.5 million light-years away. When **astronomers** look at the light coming from the Andromeda galaxy, it's as though they are looking back in time, because it has taken that light at least 2.5 million years to reach Earth.

Astronomers also use two special units to measure distances in space: **parsecs** and **astronomical units**. A parsec is equal to 3.26 light-years. An astronomical unit (AU) is equivalent to the average distance between Earth and the Sun, about 150 million kilometres.

What is a light-year?

❷ Galaxies are classified by their shape

Astronomers classify galaxies into three major groups according to their shape: elliptical, spiral and irregular (Figure 6.6).

We cannot look at our galaxy, the Milky Way, from outside it. However, by observing our night sky and comparing it with images of other galaxies, astronomers have determined that the Milky Way is a spiral galaxy. The Milky Way is part of a cluster of galaxies known as the Local Group. This includes the nearby Andromeda galaxy, the Triangulum galaxy and the Large and Small Magellanic Clouds.

What type of galaxy is the Milky Way?

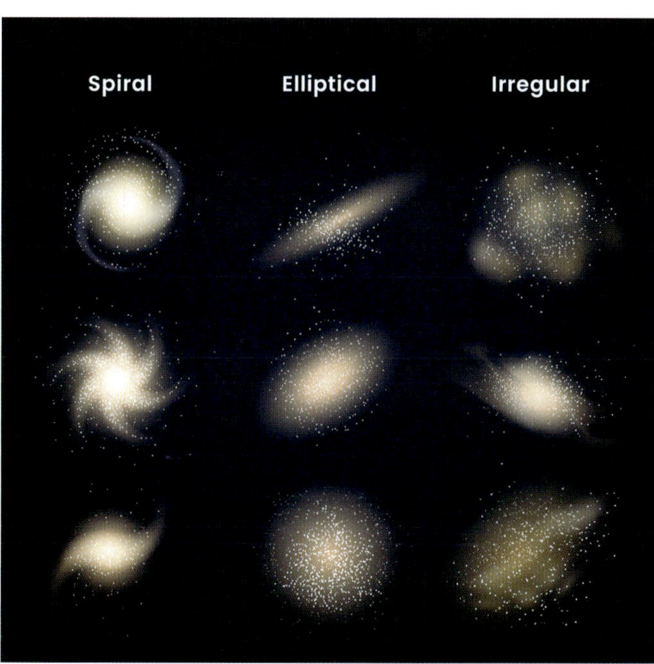

Figure 6.6 Galaxies can be classified into three main categories: spiral, elliptical and irregular.

❸ Scientists are still studying how galaxies form

Astronomers do not yet have a good understanding of how the different types of galaxies form, although they know that the forces of gravity cause stars, gas and dust to clump together. One popular theory is that irregular and elliptical galaxies form when galaxies collide and interact with one another, and that spiral galaxies form on their own, with their spiral shape resulting from the spinning motion of the galaxy.

Using telescopes, astronomers have observed that galaxies that are further away appear different from closer galaxies. This is because galaxies further away are much older, but we are observing them at a much younger stage in their life.

Evidence shows that supermassive black holes are in the centres of most galaxies. Their gravitational fields influence the stars, dust and gas in the rest of the galaxy.

What is at the centre of most galaxies?

CHECKPOINT 6.3

1 Explain how gravity is important for the formation of galaxies.

2 What evidence do astronomers use to work out how galaxies might have evolved?

3 Explain why looking at galaxies further away from us is like looking back in time.

4 What pieces of evidence have astronomers used to determine that the Milky Way is a spiral galaxy?

5 Outline some of the major features of galaxies.

EXTENSION

6 The Milky Way is moving towards the Andromeda galaxy and they are predicted to start to collide in 3.75 billion years. Predict what the shape of the new galaxy could be. Carry out research to see if your predictions match those of astronomers.

SUCCESS CRITERIA

- I can describe what a galaxy is.
- I can describe some major features of galaxies.
- I can identify the three main types of galaxies.
- I can describe what a light-year is, and how it is used to determine distances within and between galaxies.

6.4 THE SOLAR SYSTEM

LEARNING INTENTION

At the end of this lesson I will be able to outline some of the major features of the solar system and use appropriate scales to describe sizes and distances between them.

KEY TERMS

exoplanet
a planet outside our solar system

frost line
a boundary just inside Jupiter's orbit

LITERACY LINK

VOCABULARY

The word *planet* comes from a Greek word meaning 'wanderer'. Consider what observations were made by the Ancient Greeks to give planets this name.

NUMERACY LINK

CALCULATION

An astronomical unit (AU) is equal to 150 000 000 km. Use the data in Figure 6.7 to calculate the distances in kilometres between Earth and the other planets in the solar system.

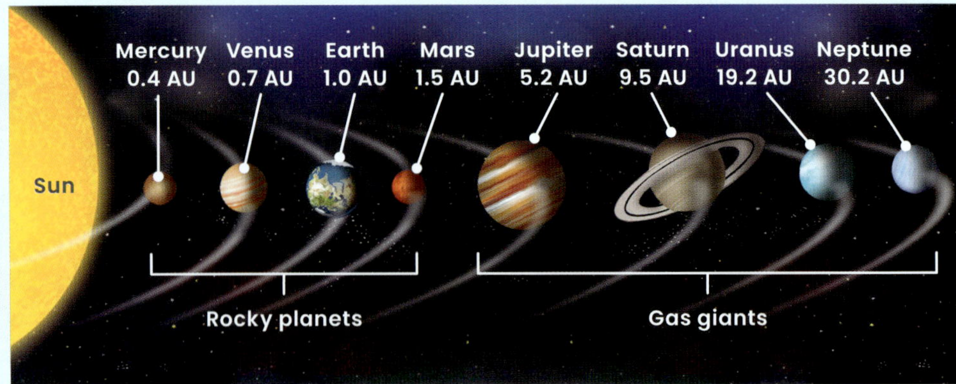

Figure 6.7 The astronomical unit (AU) is the average distance between Earth and the Sun. 1 AU = 150 000 000 km. It is useful for describing distances in the solar system.

Our solar system formed around 4.6 billion years ago. The planets formed from the dust and gas around the Sun, with the rocky planets close to the Sun, and the gas and ice giants further out. Astronomers have detected planets around other stars, and continue to search for these to learn more about our solar system.

1 The rocky planets formed from heavy elements

The nebula that our solar system formed from was mostly made up of lighter elements such as hydrogen and helium. Heavier elements such as iron, nickel, silicon and aluminium were much rarer. This is why the inner, rocky planets, such as Venus and Earth, are relatively small, because they are mostly made up of these heavier elements.

The inner solar system was too warm for substances with low boiling points (such as water and methane) to condense and exist as liquids. So, only compounds with high melting points formed, such as metals. It was here that the rocky planets formed. These high melting compounds are quite rare in the universe so the rocky planets are relatively small.

Why are the rocky planets small?

2 The gas giants mostly formed from hydrogen and helium

The **frost line** is found just inside Jupiter's orbit. Outside this boundary, temperatures are so low that hydrogen, helium and other compounds that are gases on Earth (such as methane and carbon dioxide) are able to condense and exist as liquids.

Gas giants have cores made up of rock and ice, but are otherwise mostly gases. Jupiter and Saturn, the largest planets in the solar system, are made up of large amounts of hydrogen and helium. Their large cores allowed the planets to attract lots of hydrogen and helium before the solar wind cleared the solar system. Uranus and Neptune are further out, with larger orbits.

What is the frost line?

3 Comets come from the Kuiper Belt and Oort Cloud

Comets are like dirty snowballs; chunks of rock and ice moving through space. The 'tails' that we observe are formed when they melt as they move closer to the Sun.

Short-period comets, which orbit the Sun fairly frequently, originate in the Kuiper Belt. The Kuiper Belt is an area of the outer solar system that extends from the orbit of Neptune to about 55 AU from the Sun. It is a region of icy bodies left over from the formation of the solar system. The Kuiper Belt is home to dwarf planets such as Pluto.

Long-period comets have very slow orbits; the Hale–Bopp comet orbits the Sun once every 2500 years. These comets come from the Oort Cloud, a spherical cloud of icy bodies also left over from the formation of the solar system. The Oort Cloud lies in the outermost parts of the solar system, 5000–100 000 AU from the Sun.

Where do long-period comets come from?

4 Many other solar systems have been detected

Astronomers have found many different types of solar systems. Some are similar to ours, while others are very different. Planets that orbit stars other than the Sun are called **exoplanets**. The first exoplanets were observed in 1995 and since then thousands of exoplanets have been discovered. Small rocky planets like Earth, Mars, Venus and Mercury are very common in other solar systems.

Studying other solar systems helps astronomers build a better picture of how our solar system formed. The main way that astronomers search for exoplanets is to look for a dip in the light coming from the star when the planet passes across it.

What is an exoplanet?

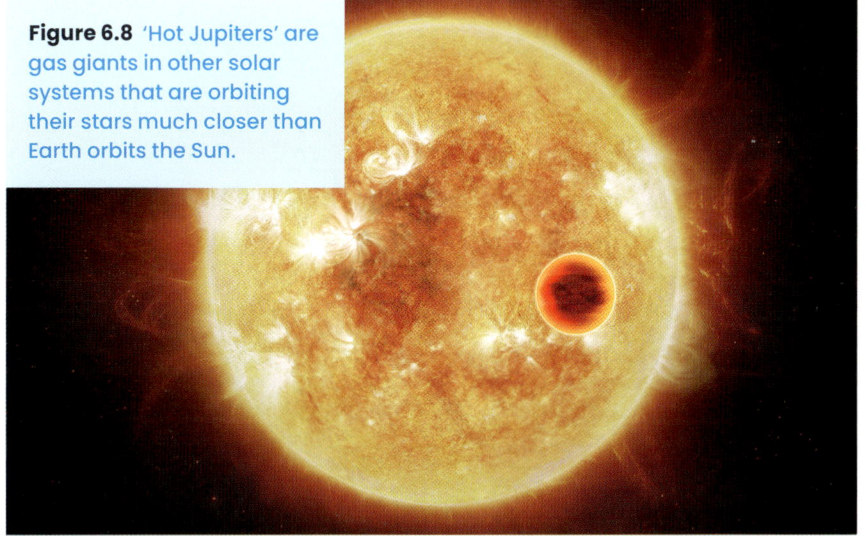

Figure 6.8 'Hot Jupiters' are gas giants in other solar systems that are orbiting their stars much closer than Earth orbits the Sun.

INVESTIGATION 6.4
Investigating orbits

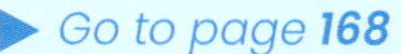

KEY SKILL
Representing data to identify patterns and trends

▶ *Go to page **168***

CHECKPOINT 6.4

1 Using your own words, outline some of the major features of our solar system.
2 Explain the significance of the frost line in the formation of the gas giants.
3 Identify at least one reason for the rocky planets being so much smaller than the gas giants.
4 Draw and label a diagram that illustrates the steps that occurred in the formation of the solar system.
5 Compare and contrast the Kuiper Belt with the Oort Cloud.
6 Describe how astronomers find exoplanets.

EXTENSION

7 Create a scale that depicts the distances between the Sun and the planets in the solar system.

SUCCESS CRITERIA

- I can describe some of the major features of the solar system.
- I can use appropriate units to describe the size of planets in our solar system and the distances between them.
- I can describe how our solar system formed.

6.5 TELESCOPES

LEARNING INTENTION

At the end of this lesson I will be able to describe how technological developments in telescopes have advanced scientific understanding about the universe.

KEY TERMS

resolution
the ability to tell two separate objects apart

LITERACY LINK

READING

In a table, summarise the information on this spread about telescope types, what they can detect and how they are useful to astronomers.

NUMERACY LINK

CALCULATION

To compare the light-gathering powers of two telescopes, you divide the area of one mirror by the area of the other. For example, a 40 cm diameter mirror on one telescope has four times the light-gathering power of a telescope with a 20 cm diameter mirror.

Using the information from the text, compare the light-gathering power of the James Webb telescope to the Hubble Space Telescope.

Telescopes have allowed us to improve our understanding about the universe: what it contains, how it formed and what our place is in it. Technologies have improved greatly from Galileo's simple telescope that allowed him to observe the Moon and planets. Telescopes can now be as large as buildings – located on remote mountain tops, and even in space. Today's telescopes can even detect more than just visible light.

1 Optical telescopes collect and amplify light

An optical telescope works by collecting and amplifying light. Galileo was the first person credited with using a telescope to study stars and the solar system. He observed four of the moons orbiting Jupiter, the rings around Saturn, and the mountains on the Moon. His refracting telescope used two glass lenses – one large lens to collect light, and a smaller eyepiece lens to magnify the image.

Today, large telescopes have mirrors to collect light, because large mirrors are easier to make than large lenses. The larger the objective lens or mirror, the more light that can be collected, and the more fine detail can be observed. This is called **resolution**. The further away an object is, the larger the telescope must be to have good resolution.

Most optical telescopes are located away from populated areas that cause light pollution. They also tend to be located at high altitudes to limit atmospheric disturbance.

Where are most optical telescopes located?

Figure 6.9 The Anglo Australian Telescope is the largest optical telescope in Australia. Scientists in Australia have led crucial research into the use of telescopes to explore and study the universe. For example, Brian Schmidt has been recognised for his work on the universe's expansion, and Penny Sackett has even discovered a planet!

2 Radio telescopes detect radio waves from excited hydrogen atoms

Hydrogen is the most common element in the universe, and when hydrogen atoms are excited by energy, they emit radio waves. Radio telescopes detect these radio waves from space, allowing astronomers to map the shape of galaxies.

Unlike light, radio waves can also travel through dust clouds, so astronomers can discover and map out objects that cannot be seen with an optical telescope. Radio astronomy has enabled the discovery and study of pulsars (rapidly spinning neutron stars), quasars (primordial galaxies with supermassive black holes), supernova remnants, and black holes in the centres of galaxies.

Radio telescopes need to be situated away from large populations where there are few radio signals from radio, television and mobile phones. Radio telescopes can be built in arrays of much smaller antennas that work together to detect faint radio signals.

What can radio telescopes be used to observe?

3 Space telescopes are located in outer space

Space telescopes allow astronomers to gather clearer images because they don't experience interference by the atmosphere or light pollution.

The Hubble Space Telescope was launched in 1990. The telescope orbits 547 km above Earth. Its main mirror has a diameter of 2.4 m, and it was able to produce images about 50% sharper than an optical telescope on Earth. The discoveries made by the Hubble Space Telescope have significantly advanced our knowledge of the universe.

The James Webb Space Telescope is due to be launched in 2021. This telescope will not orbit Earth, but will sit at a point 1 500 000 km away, orbiting the Sun. This telescope will have a mirror about 6.5 m in diameter. It will gather light in the near infrared part of the spectrum to find out more about the early universe, as well as searching in dust clouds for stars and solar systems that are forming.

What is the advantage of a space telescope over a ground-based telescope?

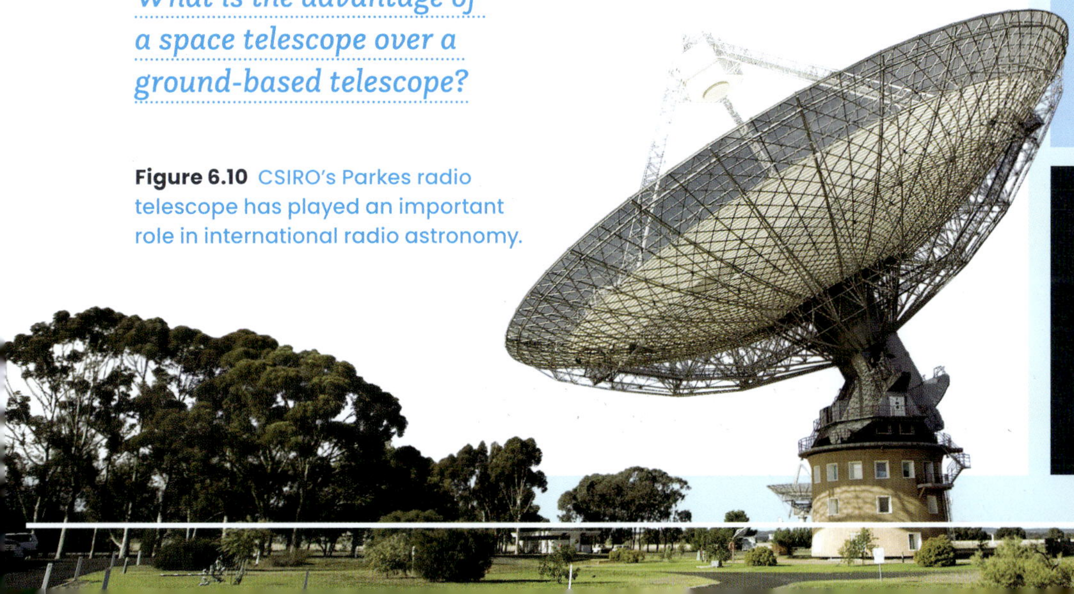

Figure 6.10 CSIRO's Parkes radio telescope has played an important role in international radio astronomy.

INVESTIGATION 6.5A
Making a telescope

KEY SKILL
Writing a research question
Go to page **170**

INVESTIGATION 6.5B
Lens diameter and resolution

KEY SKILL
Representing data to identify patterns and trends
Go to page **171**

CHECKPOINT 6.5

1 Copy and complete.
Resolution is the ability to distinguish between ________ separate objects. The further away the objects, the ________ the telescope needs to be.

2 Explain why mirrors are used rather than lenses in large telescopes.

3 Why would using an array of radio telescopes be an advantage?

4 Some telescopes can stay focused on an object as Earth rotates, allowing more images to be gathered. Explain the advantage this would have over a telescope that does not move.

RESEARCH

5 Research the contribution of either astronomer Penny Sackett or astrophysicist Brian Schmidt to the field of space science. Summarise your research in a one-page report.

SUCCESS CRITERIA

- I can describe optical telescopes, radio telescopes and space telescopes.
- I can describe the advantages of a space telescope.

VISUAL SUMMARY

◀ Developments in technology have advanced scientific understanding of the universe.

Galaxies are groups of stars, gas and dust bound together by gravity.

Black holes form when supermassive stars die.

Stars are mostly hydrogen.

1 H Hydrogen 1.0079

Hydrogen is the most common element in the universe.

Galaxies are classified by their shape.

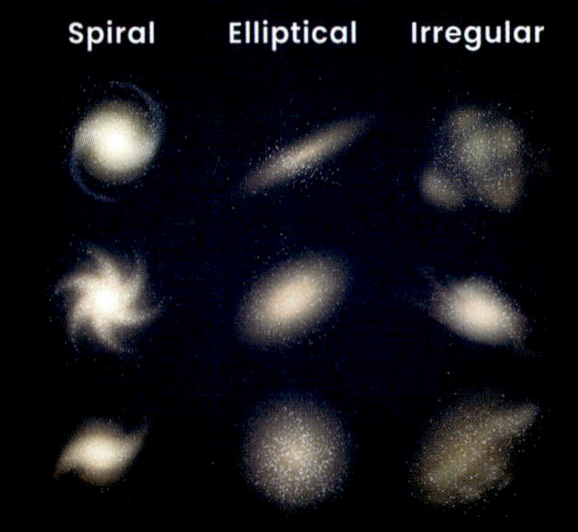

The big bang theory is the currently accepted theory for the origin of the universe.

1 AU = 150 000 000 km = distance from the Sun to Earth

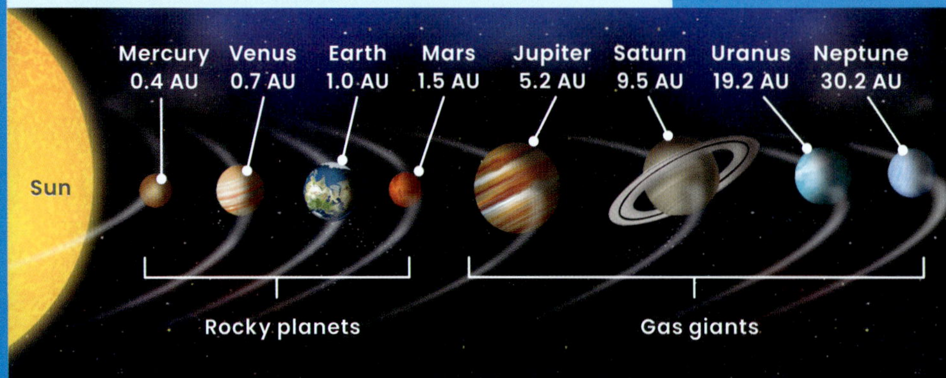

Evidence for the big bang theory includes:

- the universe is expanding
- there is an abundance of light elements
- cosmic microwave background radiation.

FINAL CHALLENGE

1 Explain the big bang theory in your own words.

2 Describe what stars and galaxies are made of.

3 A galaxy is best defined as:

- a a region where stars are forming.
- b a cloud of dust and gas.
- c a mass of stars, gas and dust bound together by gravity.
- d the gas left over after a supernova.

Level 1 50xp LEVEL UP!

4 The Sun will end its life as a:

- a red dwarf.
- b white dwarf.
- c planetary nebula.
- d supernova.

Level 2 100xp LEVEL UP!

5 What is a light-year and why is it used to determine distances within and between galaxies?

6 Identify the three main shapes used to classify galaxies.

7 Explain why the rocky planets are close to the Sun and the gas giants are further away.

8 Identify an appropriate situation to use distances measured in:

- a astronomical units
- b light-years.

Level 3 150xp LEVEL UP!

9 Outline what evidence exists for the age of the universe.

10 Describe the various ways that stars' lives can end.

11 Describe the advantages of space telescopes over ground-based telescopes.

Level 4 200xp LEVEL UP!

12 Identify and explain three pieces of evidence that support the big bang theory of the origin of the universe.

13 Discuss this statement using evidence to support your response.
'Our understanding of the universe cannot improve unless technology to observe the universe also improves.'

ENERGY

7

How can increased understanding of energy and motion change the world?

When the universe formed, so too did all the energy there is and ever will be. No new energy is ever created or destroyed; it is just transferred or transformed, cycling around the universe, forever.

Most of the energy to power your house comes from burning fossil fuels such as coal. But mining and burning fossil fuels has serious impacts on the environment, such as climate change, habitat loss and air pollution.

1 FRAYER MODEL

Copy and complete the below chart in your workbook.

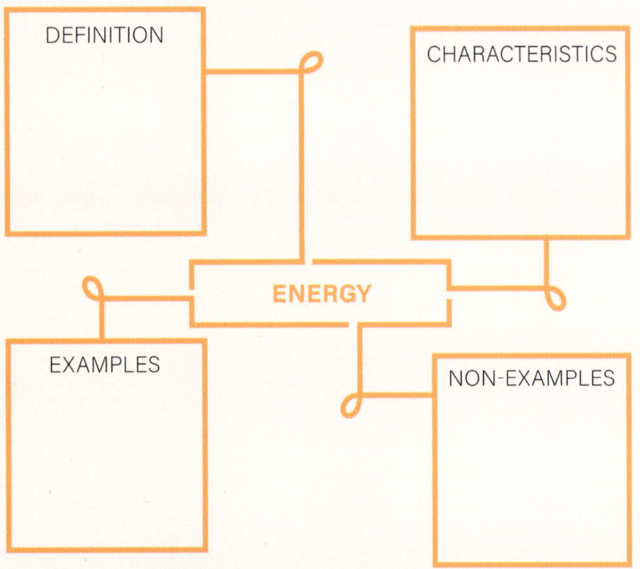

Complete two additional charts for the key terms *Waste* and *Renewable*.

2 LEARNING LINKS

Brainstorm everything you already know about energy.

3 SEE-KNOW-WONDER

List three things you can SEE, three things you KNOW and three things you WONDER about this image.

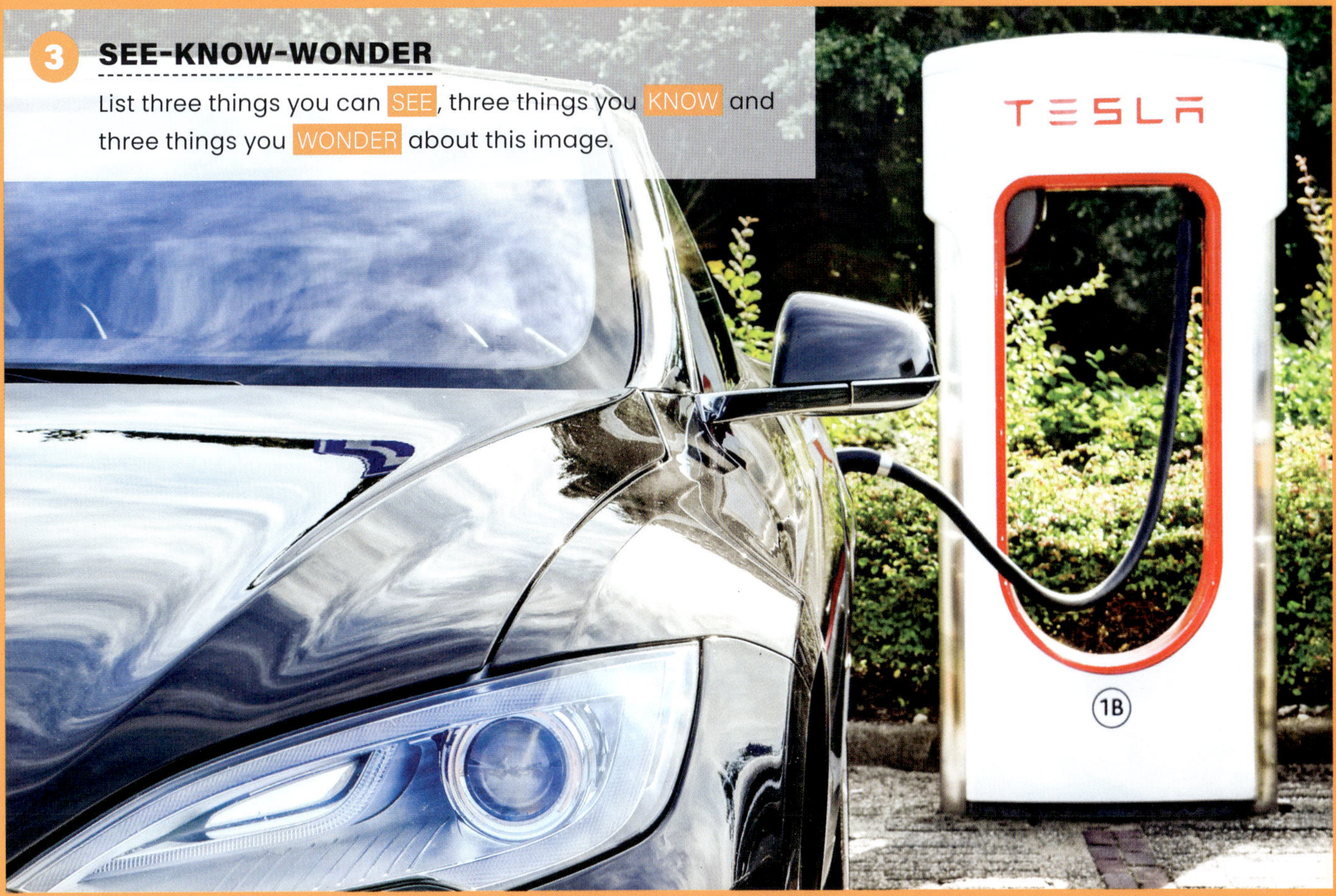

4 CRITICAL + CREATIVE THINKING

MISMATCH: How could you stop a speeding arrow with a sheet of paper, a dog collar and a cowboy boot?

THE QUESTION: Write five questions that have the answer 'energy'.

THE RIDICULOUS! Attempt to support this statement: 'We should burn forests instead of coal to generate electricity.'

5 THE BEST INTENTIONS!

In the 1920s, Thomas Midgley discovered that adding lead to petrol improved engine performance, which led to the popularity of leaded petrol. However, we now know that leaded petrol is harmful to the environment and people, so leaded petrol is banned. Midgley also helped develop CFCs, chemicals used in refrigerators, which caused a giant hole in the ozone layer. So that's two of Midgley's inventions that had devastating worldwide impacts!

7.1 CONSERVATION OF ENERGY

LEARNING INTENTION

At the end of this lesson I will be able to recognise that the law of conservation of energy explains that total energy is maintained in energy transfers and transformations.

KEY TERMS

energy
a measure of the ability to do work

energy transfer
the movement of energy from one place to another without changing form

energy transformation
a change from one type of energy to another

LITERACY LINK

VOCABULARY

Describe five types of energy mentioned in this section in ways that a primary school student could understand.

NUMERACY LINK

GRAPHING

Sketch a line graph of energy versus time, showing how the kinetic energy and gravitational potential energy of a ball change over time as it bounces on a hard surface.

Many of the devices you use every day require **energy**. However, this energy may need to be transferred from one place to another and transformed from one type of energy into another.

For example, a toaster will heat up to toast bread. For this to happen, the electrical energy needs to be transferred from the power station that produces it to the toaster, where it then needs to be transformed into heat energy. These types of transfers and transformations of energy are crucial for a countless number of systems to be able to operate effectively.

1 Energy cannot be created or destroyed

The conservation of energy is a fundamental law of physics. The law states that energy can never be created or destroyed. This means that all the energy we use now existed in some form billions of years ago and has cycled around continuously. Although energy cannot be created or destroyed, it can exist in different forms. Table 7.1 lists some types of energy.

Table 7.1 Types of energy

Type of energy	Description
Chemical	Energy stored in chemical bonds
Elastic potential	Energy stored in an object that has been stretched or compressed
Electrical	Energy that travels through electrical circuits
Gravitational potential	Energy stored in an object that has been raised above the ground
Heat	Energy that can raise the temperature of an object
Kinetic	Energy that any moving object has
Light	Energy that you can see, emitted from glowing objects
Nuclear	Energy stored in the nucleus of atoms
Sound	Energy that you can hear, caused by vibrating air particles

What is the law of conservation of energy?

Figure 7.1 Elastic potential energy is stored in a drawn bowstring.

2 Energy can be transferred from one place to another

Energy can be transferred from one location to another. A rolling marble will have kinetic energy and, if it strikes a stationary marble, both marbles will move away from the collision. There has been an **energy transfer** from the first marble to the second marble.

Another example is electrical energy, which is transferred from a power station through power lines to your house, where it comes out of a power point and into appliances. Heat energy can also be transferred from a hot stove into your finger, but that would be pretty painful!

How is electrical energy transferred from place to place?

3 Energy can change from one form to another

Energy can transform, or change, from one form to another. This can be done using technology, such as a toaster turning electricity into heat, but it also happens naturally. When you throw a ball up in the air, the ball starts off moving quickly, slows down as it gets higher, and then accelerates back to the ground. The ball starts off with a lot of kinetic (movement) energy. As it gains height, the ball loses speed. An **energy transformation** takes place as kinetic energy is converted to gravitational potential energy.

Figure 7.2 A toaster converts electrical energy into heat energy.

When thrown straight up into the air, at its peak, the ball briefly stops – it has no kinetic energy, but its gravitational potential energy is at a maximum. As the ball comes back down, the gravitational potential energy is transformed to kinetic energy, making the ball travel faster and faster as it loses height.

As a ball is thrown upwards, what type of energy is being converted?

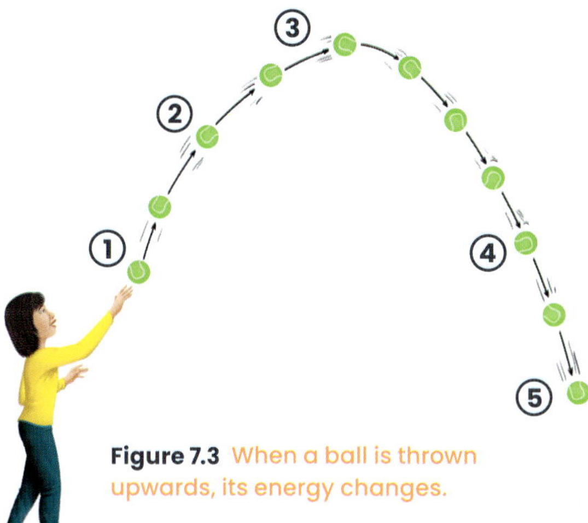

Figure 7.3 When a ball is thrown upwards, its energy changes.

1 The ball starts with a lot of kinetic energy and moves fast.
2 It slows down as its kinetic energy is converted to gravitational potential energy.
3 The ball stops for an instant; it has no kinetic energy.
4 It speeds up as its potential energy is converted into kinetic energy.
5 The kinetic energy is transferred to the ground and is converted to sound and heat energy.

INVESTIGATION 7.1
Galileo's pendulum

KEY SKILL
Representing and recording data using a table

▶ *Go to page **172***

CHECKPOINT 7.1

1 Describe the law of conservation of energy.
2 Explain the difference between energy transfer and energy transformation.
3 What type of energy transformation is happening when a:
  a power station burns coal to provide household power?
  b pot of water is placed on a gas stove to boil?
  c slingshot is stretched back and then released, firing a marble?
4 What kind of device could transform:
  a electrical energy into light and sound energy?
  b chemical energy into heat energy?
  c light energy into chemical energy?

CRITICAL AND CREATIVE THINKING

5 Design a vehicle that converts elastic potential energy into kinetic energy. Make sure you explain how the elastic energy is stored in the first place.

SUCCESS CRITERIA

- I can explain the law of conservation of energy.
- I can describe what is meant by energy transfers and transformations.

7.2 WASTED ENERGY

LEARNING INTENTION

At the end of this lesson I will be able to recognise that in energy transfers and transformations, a number of steps can occur and the system is not completely efficient, so usable energy is reduced.

KEY TERMS

efficient
not wasteful

energy efficiency
how much usable energy is produced compared to how much energy has been supplied

friction
a contact force that opposes motion, caused by objects rubbing against each other

joule
the unit of energy

usable energy
the type of energy that a device is designed to produce

LITERACY LINK

READING

After reading this lesson, write five questions about wasted energy that you would like to have answered.

NUMERACY LINK

UNITS

A TV that is 86% efficient produces 5300 J of useful energy.

a How much energy does the TV consume?

b Convert your answer to kJ.

Energy can exist in many different forms, and many mechanical devices simply convert one type of energy to another. A device is usually designed to produce a particular type of energy, known as usable energy.

However, devices inevitably produce other types of energy that is not used. This energy is wasted. For example, a light bulb is designed to produce light energy, but it also produces some heat.

Figure 7.4 When a light bulb produces light, it also produces some heat.

1 Energy is lost in energy transfers and transformations

Consider a tennis ball falling from a height. The ball starts with gravitational potential energy. As the ball falls, the gravitational potential energy is converted to kinetic energy. When the ball strikes the ground, it compresses and stores elastic potential energy. The elastic potential energy is then transformed into kinetic energy as the ball bounces back up.

The ball never bounces back to its original height even though energy can't be created and destroyed. This is because the ball is losing energy in a few places.

- As the ball is falling, it experiences air resistance, a form of **friction**, with the air around it. This causes the ball to heat up very slightly and lose some energy as heat.
- As the ball strikes the ground, it makes a sound. Some of its kinetic energy is converted to sound energy.
- As the ball strikes the ground, a small amount of heat energy is transferred to the ground.

These factors all contribute to a loss of energy, so the ball finishes with less energy than it started with.

What types of energy are lost as a ball bounces?

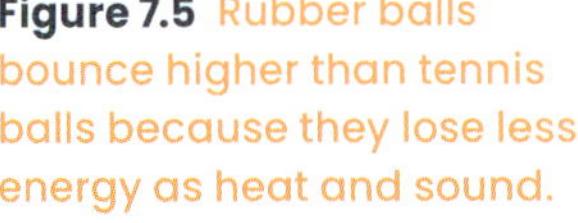

Figure 7.5 Rubber balls bounce higher than tennis balls because they lose less energy as heat and sound.

❷ Most devices lose energy

Devices are designed to produce **usable energy**. Almost all devices lose energy. After riding your bicycle, your tyres will have warmed up. This is because of the friction between the tyres and the road. This 'lost' energy has been converted from usable kinetic energy into wasted heat energy.

Moving objects often lose some energy as heat, because of friction. A ball won't keep rolling forever because its kinetic energy is slowly converted to heat as a result of friction with the floor and the air. Friction can be helpful in some situations. For example, brakes in vehicles work by using friction to convert kinetic energy into heat – otherwise we'd never be able to stop!

Energy in devices can also be lost in other ways. A car loses energy through engine noise and heat. Most electrical devices in your home lose energy as heat and sound (even if the sound is just a quiet buzzing).

What causes bicycle tyres to heat up during a ride?

Figure 7.6 A spacecraft re-entering the atmosphere experiences friction with the air.

❸ All devices are less than 100% energy efficient

Devices are often described as energy **efficient**. This means that they don't waste very much energy. The **energy efficiency** of a device is expressed as a percentage, calculated by dividing the amount of usable energy produced by the amount of energy put into the device. The unit of energy is the **joule** (J).

$$\text{Energy efficiency (\%)} = \frac{\text{energy output}}{\text{energy input}} \times 100$$

When a smartphone is fully charged, the energy in the battery should be converted to light energy (to show things on the screen) and sound energy (for phone calls or listening to music). If the phone heats up a lot while using it, some of the stored energy is lost as heat energy, leaving less energy to power the phone, resulting in the battery going flat earlier.

If a smartphone were 40% efficient, then for every 100 J of energy in the battery, only 40 J would be turned into light and sound, and 60 J would be lost as heat. If a smartphone were 80% efficient, it would only waste 20 J as heat and so the battery would last much longer.

However, almost all devices have an efficiency of less than 100%.

How is energy efficiency calculated?

INVESTIGATION 7.2

Energy efficiency of bouncing balls

KEY SKILL
Identifying the variables and formulating a hypothesis

▶ Go to page **173**

CHECKPOINT 7.2

1 What does it mean if a device is 'energy efficient'?

2 Wasted energy is usually what type of energy?

3 Friction is often the reason that kinetic energy turns into heat energy. Give an example where this is a problem, and an example where this is useful.

4 What types of usable energy do smartphones produce?

5 Explain, in terms of energy, why a cannonball fired from a cannon slows down as it travels.

6 For every 100 J of electrical energy put into a kettle, 30 J is lost as unwanted sound and heat energy. What is the percentage efficiency of the kettle?

CRITICAL AND CREATIVE THINKING

7 The refrigerator is one of the most energy-inefficient appliances in most homes. Design a new energy-efficient refrigerator.

SUCCESS CRITERIA

- I can describe how energy is lost in energy transfers and transformations.
- I can explain what is meant by energy efficiency.

7.3 THE IMPORTANCE OF ENERGY EFFICIENCY

LEARNING INTENTION

At the end of this lesson I will be able to discuss, using examples, how society's values and needs influence scientific research into increasing the efficiency of electricity use by individuals and society.

KEY TERMS

finite
limited; won't last forever

fossil fuel
a fuel that is formed from the decomposition of dead animals and plants over millions of years; e.g. oil and coal

generate
to produce or make something

LITERACY LINK

SPEAKING

Write a short speech encouraging people to convert to LED light bulbs. Test your speech out on a partner, then listen to theirs.

NUMERACY LINK

DATA

Kat is testing some light bulbs to determine their power usage. She records the power usage of five bulbs to be 94 W, 48 W, 23 W, 65 W and 7 W. She concludes that the 7W bulb is the best. Do you agree? What other information would you need to make a stronger conclusion?

Imagine you lived in a city where everyone was constantly getting sick. You would probably want the scientists in your city to investigate why people were getting sick and how to cure the sickness.

Scientists are often employed to solve problems that a particular society is facing. In recent times, people have begun to realise that fossil fuels won't last forever, and that using them produces pollution and is contributing to climate change. People have looked to scientists to determine how to make fossil fuels last as long as possible, and to research potential replacements.

1 Most of our energy is generated by burning fossil fuels

Most of Australia's power and electricity is **generated** (produced) by burning fossil fuels. **Fossil fuels** include coal and oil, which take millions of years to form. In particular, Australia's electricity is mostly generated by burning coal.

Unfortunately, these resources are **finite** and will run out one day. Also, burning fossil fuels emits harmful greenhouse gases into the atmosphere. This means that eventually we will need to find other ways to generate electricity. In the meantime, it is vitally important to minimise the amount of energy we waste. For example, imagine you have the choice between two light bulbs that both produce 8 J of light energy every second. One of them is 80% efficient and the other is only 50% efficient. The first one uses 10 J of energy, but the second needs 16 J of energy for the same light output. Which will you choose?

Where does most of Australia's electricity come from?

Figure 7.7 Fossil fuels such as coal currently power most of Australia's electricity grid.

❷ Energy-efficient devices are becoming more popular

As energy prices increase and we understand more about the problems arising from using fossil fuels, more people are interested in buying energy-efficient devices. Energy efficiency is becoming a popular feature of appliances for a number of reasons, including lower running costs and the benefit to the environment.

In addition, many governments around the world are interested in energy efficiency, with some enacting laws requiring minimum standards. This has meant that a lot of scientific research has been directed at improving the energy efficiency of appliances and systems.

What are some benefits of energy-efficient devices?

❸ Light bulbs are an example of increasing energy efficiency

One very obvious example of society driving a change to more energy-efficient products has been with light bulbs. Old incandescent light bulbs are very inefficient, wasting 90% of their energy as heat. Halogen lamps are about 10–20% more efficient than incandescent bulbs, so governments around the world started phasing out the use of the less efficient bulbs in the mid-2000s. Compact fluorescent bulbs and LED lights use even less energy, and the Australian government began phasing out halogen bulbs in 2020.

By phasing out these older bulb types, people will pay less to light their homes and will use less energy, reducing the amount of fossil fuel we need to burn.

What is the least efficient type of light bulb?

Figure 7.8 Light bulbs have changed over time.

CHECKPOINT 7.3

1 Describe some disadvantages of using fossil fuels to generate electricity.

2 Suggest why energy-efficient devices are better for the environment.

3 Give two reasons why society is moving towards energy-efficient devices.

4 Suppose a company that makes toasters refused to develop more energy-efficient models. What do you think would happen to this company?

5 a List these bulb types in order from most to least efficient: compact fluroescent, incandescent, LED, halogen.
  b Which of the bulb types in part a would you expect to get the hottest as it runs? Explain your answer.

6 Why do you think the government started phasing out inefficient light bulbs?

RESEARCH

7 Australia uses a star rating system to classify the energy efficiency of appliances. Find out what these stars mean and try to find three appliances with star ratings in your house.

SUCCESS CRITERIA

- I can explain why energy-efficient devices are important.
- I can describe how the needs of society can drive scientific research.

7.4 ENERGY SOLUTIONS

LEARNING INTENTION

At the end of this lesson I will be able to research ways in which scientific knowledge and technological developments have led to finding a solution to a contemporary issue; e.g. improvements in devices to increase the efficiency of energy transfers or conversions.

KEY TERMS

insulation
material that acts as a barrier to heat flow

passive design
building design that considers climate and location to make the best use of natural heating and cooling

thermal
to do with heat

LITERACY LINK

LISTENING

Verbally summarise one of the sections in this lesson to a partner. Your partner has to think of 'two stars and a wish' – two things you did well and one thing that could be improved. Then swap roles, and listen to your partner's summary.

NUMERACY LINK

CALCULATION

A solar power system costs $5000 to install, but saves the homeowner $750 per year. How long will it take until the cost of the system is paid back in full?

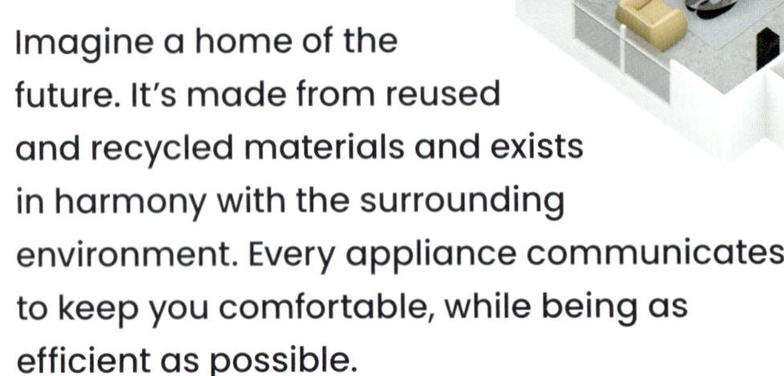

Figure 7.9 In the future, homes and workplaces will need to be more energy efficient.

Imagine a home of the future. It's made from reused and recycled materials and exists in harmony with the surrounding environment. Every appliance communicates to keep you comfortable, while being as efficient as possible.

Would you like to live there? Let's find out how you can.

1 Passive design helps regulate a building's temperature

Passive design is building design that considers the climate and where a building will be located to keep it at a comfortable temperature without using appliances for heating or cooling.

A building made with passive design has features such as:

- windows and doors that are placed to allow good air flow
- few windows facing the hot afternoon sun
- high and low vents so that cool, fresh air replaces hot, stale air during summer and reduces heat loss during winter
- roof angles that maximise shade during summer and heat absorption during winter
- plants that give shade and cool the air, like the canopy in a rainforest.

These features can reduce or eliminate the need for heating and cooling that relies on electricity.

What does passive design aim to do?

2 Building materials can help control temperatures

Insulation acts as a barrier to heat flow. It keeps homes warm in winter and cool in summer. Insulation is rated using a system called R-values – materials with higher R-values give better insulation.

There are two different types of insulation: bulk and reflective. Bulk insulation relies on air trapped in fibres to slow the transfer of heat energy, such as when you wrap yourself in a blanket to keep warm.

Bulk insulation can be made of glass wool, cellulose fibre, polyester, wool or polystyrene. Reflective insulation uses its shiny surface to reflect heat energy away.

The colour of building materials also affects how much heat energy a building absorbs. Dark, dull surfaces absorb more heat energy than light, shiny surfaces.

What type of building material acts as a barrier to heat flow?

3 Energy sources should be renewable and used efficiently

A lot of the energy used in Australian households is electrical energy. If household electricity came mostly from renewable sources, demands on non-renewable energy sources could be reduced.

Solar energy is a common renewable energy source in Australia. Solar **thermal** systems can be used with hot water systems, and they transfer the Sun's heat energy to water. Solar power cells absorb the Sun's energy and transform it to electrical energy. This energy can be stored in batteries for use at night. Solar cell technology is becoming more efficient and affordable.

Light bulbs, washing machines and refrigerators are examples of appliances that need electrical energy to function.

Energy-efficient appliances function using minimum energy. Three ways to reduce the energy use in your house are:

- avoiding unnecessary use of appliances
- choosing energy-efficient appliances
- switching appliances off when not in use.

What is a common renewable energy source in Australia?

Figure 7.10 This Australian house uses passive design, such as the design of its roof, to help control the temperature inside.

INVESTIGATION 7.4

Insulation and heat transfer

KEY SKILL

Identifying limitations to the method and suggesting improvements

▶ *Go to page* ***174***

CHECKPOINT 7.4

1 Copy and complete the following sentences.

a Passive design is ______________.

b Household solar cells can ______________.

c Choosing energy-efficient appliances is important because ______________.

d Improved building materials allow ______________.

e Insulation is used in buildings to ______________.

2 James says he designed his home with white roofing because it will help keep the building cool. Xing says she chose black roof tiles because this colour stops the heat passing through the tiles. What do you think?

INQUIRY

3 Look around your home and make a list of any energy-saving features. Make another list of improvements that could be added to your home to save energy.

SUCCESS CRITERIA

- I can describe some ways that homes can be more energy efficient.
- I can make suggestions about improving energy efficiency in buildings, devices or other technology.

7.5 HEAT ENERGY AND THE ENVIRONMENT

LEARNING INTENTION

At the end of this lesson I will understand how energy flow in Earth's atmosphere can be explained by the processes of heat transfer.

KEY TERMS

circulate
to move freely or continuously through an area

smog
a form of severe air pollution

temperature inversion
weather conditions that stop fresh air from circulating

LITERACY LINK

WRITING

Write a letter to the editor warning about the negative effects of heat pollution.

NUMERACY LINK

GRAPHING

Coulson measures the air temperature in his neighbourhood over the course of an autumn day.

It climbs from 9°C at 9am to 15°C at 3 pm, falls to 10°C by 6 pm, and is 8°C when Coulson takes his final measurement at 9 pm.

Draw a line graph to represent the temperature changes over the day.

The energy we use at home and in factories often generates waste heat energy. This can worsen pollution, which affects air quality.

To reduce pollution, we need to use energy wisely and generate energy that doesn't waste a lot of heat energy.

1 Warm pollutants can affect air quality

Under normal conditions, the Sun's energy heats the ground, which then heats the air just above the ground. This warmer air rises as cooler air sinks below it. In this way, fresh air can **circulate**.

During a **temperature inversion**, the layer of air close to the ground is cooler than the air above it, which prevents fresh air from circulating. The warm air is sometimes a result of pollutants such as those from car exhausts and factories, which can make **smog** in the air. This inversion can be made worse in a place such as a valley, where air circulation is prevented even more by the surrounding hills.

Figure 7.11 In normal conditions (left), air is able to circulate. In a temperature inversion (right), warm air traps cooler air, preventing circulation.

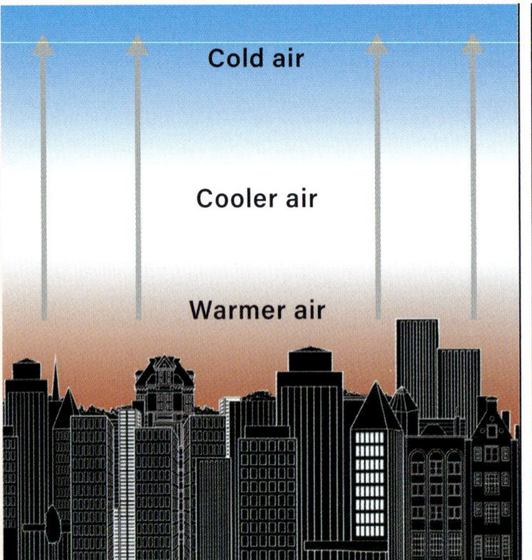

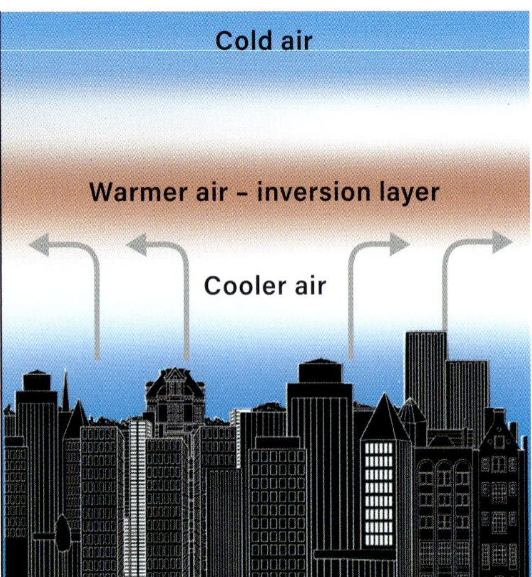

Smog contains a substance called ozone, which is a molecule made up of three oxygen atoms. Increased ozone levels can be bad for health. Breathing in smog can cause problems such as:

- difficulty breathing and lung damage
- coughing and throat irritation
- asthma symptoms.

Children, elderly people and those with respiratory conditions are at most risk of health problems from smog.

What can cause smog to be trapped around cities?

❷ Reducing heat energy waste has many advantages

There are two main ways to reduce heat energy waste in industry and at home. The first is to use improved insulation materials. Better insulation reduces heat loss, so less energy is required to heat buildings. The other way is to recover 'waste' heat using heat pumps or steam recyclers. This waste energy can be used to keep us warm, or be recirculated to avoid release into the environment.

Reducing heat energy waste helps protect our health and the environment. But there are advantages for industries too, such as saving money on energy and repairs, and having lower emissions of other pollutants.

What are some advantages of reducing pollution?

❸ Renewable energy sources make less heat

The electrical energy we use is generated in a variety of ways. It can be made from non-renewable and renewable energy resources.

Non-renewable energy resources such as coal, oil and natural gas are burned to release heat energy. This heat energy is transformed to electrical energy in fuel-fired power stations. A lot of this heat energy leaves the system in the process.

Renewable energy sources, such as wind power, hydro-electricity and solar power, don't make heat as burning fossil fuels does. This is because these energy sources can directly turn turbines to generate electrical energy, rather than heating water to steam to turn turbines.

Which energy sources produce minimal waste heat energy?

Figure 7.12 When coal is burnt in a power plant, the heat is used to boil water that turns a turbine.

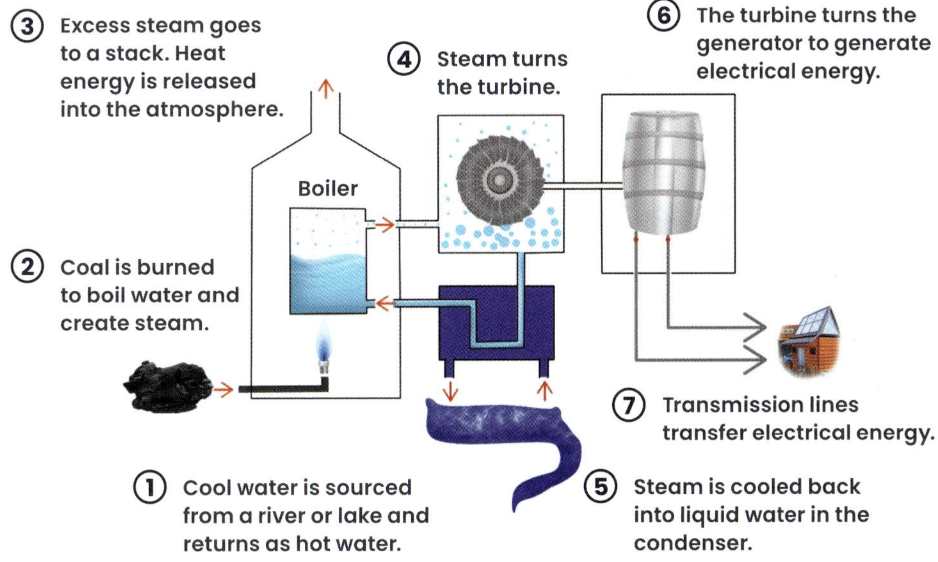

INVESTIGATION 7.5

Atmospheric convection and the inversion layer (Teacher demonstration)

▶ Go to page 175

CHECKPOINT 7.5

1 Explain what heat energy waste is and how it is generated.

2 Describe some health issues caused by smog.

3 True or false? Write the correct statement if a statement is false.
   a Wind, hydro-electricity and solar energy are renewable sources of energy.
   b Waste heat energy can't be reused.
   c An atmospheric inversion layer can trap polluted cold air around cities.

4 Andrea wants to buy an electric car because she thinks it would help reduce the use of fossil fuels. Gabriel disagrees. Who do you think is right? Justify your choice.

PERSONAL AND SOCIAL CAPABILITY

5 Brainstorm all the different groups that are either affected by or a cause of poor air quality. Consider each of their viewpoints before summarising them in a short report.

SUCCESS CRITERIA

- I can describe a mechanism of heat transfer in the atmosphere.
- I can explain how human activity influences thermal pollution.

VISUAL SUMMARY

Energy exists in many forms, such as kinetic, gravitational potential, elastic potential, heat, sound and light.

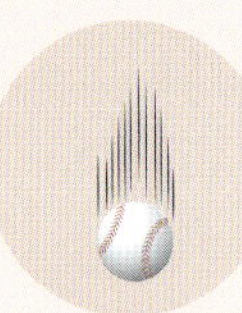

Gravitational potential energy

Electrical energy

Kinetic energy

Nuclear energy

Chemical energy

Heat energy

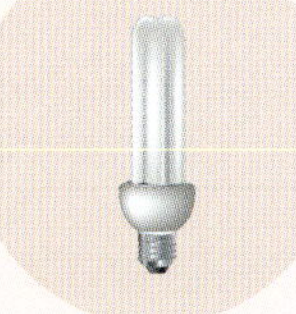

Light energy

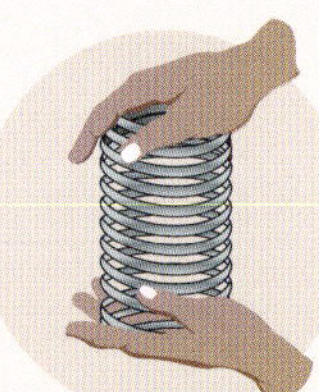

Elastic potential energy

Sound energy

The **law of conservation of energy** states that energy cannot be created or destroyed; it can only change form.

Energy can change form. Plants can convert heat and light energy into chemical energy, while a speaker converts electrical energy to sound energy.

All devices are less than 100% energy efficient.

Friction produces heat and is the most common cause of energy inefficiency:

$$\text{Energy efficiency (\%)} = \frac{\text{energy output}}{\text{energy input}} \times 100$$

Most of Australia's electricity currently comes from burning non-renewable fossil fuels.

Burning coal has negative environmental effects, such as pollution and global warming. We only have a limited amount of coal left.

The benefits of burning coal are mainly economic.

★ FINAL CHALLENGE ★

1. State the law of conservation of energy.
2. Name five different types of energy, and give examples of where they can be found.
3. Old-fashioned incandescent light bulbs are inefficient because they mostly give out what type of energy?

Level 1 ★☆☆ ☆☆☆ 50xp LEVEL UP!

4. Give one example of transferring energy and one example of transforming energy.
5. What is the source of most of Australia's electricity?
6. Into what types of energy do the following devices convert electrical energy?
   a speaker
   b television
   c kettle

Level 2 ★★☆ ☆☆☆ 100xp LEVEL UP!

7. Give three examples of situations where one type of energy is converted into another.
8. Explain why car tyres heat up after driving.
9. Describe why energy-efficient devices are better for the environment.

Level 3 ★★★ ☆☆☆ 150xp LEVEL UP!

10. If energy is always conserved, explain why a ball never returns to its full height after bouncing.
11. What is meant by 'energy efficiency'?
12. If light bulb A has an efficiency of 25%, and light bulb B has an efficiency of 40%, which one would be more expensive to run? Explain why.

Level 4 ★★★ ★☆☆ 200xp LEVEL UP!

13. Should we continue to use fossil fuels to generate electricity? Explain your point of view.
14. Imagine you are tasked with battling climate change in Australia. Create a plan that lists five recommendations in order of importance.
15. Create a flow chart that shows the loss of energy from transfers and transformations involved in running a vacuum cleaner.

Level 5 ★★★ ★★★ 300xp LEVEL UP!

8 MOTION

How can increased understanding of energy and motion change the world?

How coordinated are you? Do you find it easy or hard to catch a ball? When you catch a ball, your brain does some very rapid calculations so that you can quickly work out the ball's speed, angle and position. The motion of cars, satellites and bullets is more complex than the motion of a thrown ball. However, you can still predict how complex objects will move by using the laws of motion. These laws govern how all sorts of things move under different conditions and forces.

1 FRAYER MODEL

Copy and complete the below chart in your workbook.

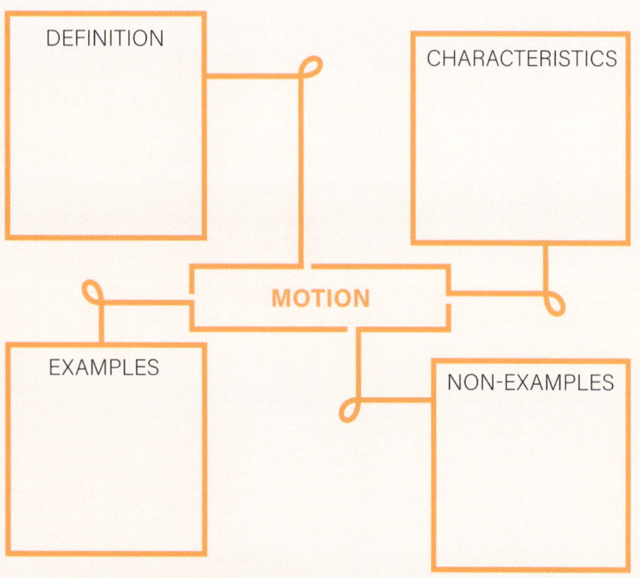

Complete two additional charts for the key terms *Force* and *Acceleration*.

2 LEARNING LINKS

Brainstorm everything you already know about motion.

3 SEE-KNOW-WONDER

List three things you can SEE, three things you KNOW and three things you WONDER about this image.

CRITICAL + CREATIVE THINKING

WHAT IF ... Earth and the Moon suddenly swapped gravitational fields?

10 THINGS: List 10 things that cannot be accelerated.

ALTERNATIVES: List all the ways you could change the speed or direction of a car without touching the pedals or steering wheel.

THE STRONGEST!

The strongest force in the universe is called the strong nuclear force. This force keeps the smallest particles in an atom (called quarks) stuck together. Luckily, the strong nuclear force doesn't work over distances greater than the nucleus of an atom. Otherwise, the force of attraction between all matter would be so great that everything on Earth would be squashed together into the size of a house, and a tennis ball would weigh as much as 3 million Suns!

8.1 FORCE, MASS AND ACCELERATION

LEARNING INTENTION

At the end of this lesson I will be able to describe the relationship between force, mass and acceleration.

KEY TERMS

acceleration
a change in speed or direction of motion over time

force
a push, pull or twist

inversely proportional
when one value decreases at the same rate that the other increases

mass
the amount of matter in an object

newton
the unit of measurement of force

LITERACY LINK

WRITING

Write a short paragraph to explain the relationship between force, mass and acceleration to a primary school student.

NUMERACY LINK

CALCULATION

Applying a force of 1440 N to a car sees it accelerate by 1.2 ms^{-2}. Calculate the mass of the car.

Have you ever tried to push a car? It requires a lot of force to get a car moving. On the other hand, you can get a skateboard moving with just your foot. The difference is mass: a car is much heavier than a skateboard, and so requires a much greater force to make it accelerate.

This tells us that there must be a relationship between force, mass and acceleration.

Figure 8.1 A golf ball can be accelerated rapidly because it has a small mass.

1 Acceleration depends on mass and the size of the force being applied

The relationship between force, mass and acceleration is summed up by Newton's second law, which says that the acceleration of an object is proportional to the force applied, and **inversely proportional** to the mass of the object.

Newton's second law can also be written as an equation:

$$\text{force} = \text{mass} \times \text{acceleration}$$

If you increase the force on an object of fixed mass, the acceleration will increase. If you use the same force but increase the mass, the acceleration will decrease.

Acceleration is a measure of how quickly an object's speed or direction of motion changes. Acceleration is usually measured in metres per second squared (ms^{-2}). If you drop a ball, it accelerates towards the ground – that is, its speed increases downwards. If you press the accelerator pedal in a car, the car's speed increases in the direction it is moving.

As long as speed or direction is changing, the object is accelerating. However, if an object isn't moving or is moving at a constant speed, then it is not accelerating.

What is acceleration a measure of?

Figure 8.2 Drag cars undergo rapid acceleration, and so their speed increases very quickly.

❷ Mass is the amount of matter in an object

The **mass** of an object is a measure of how much matter it contains and is usually measured in kilograms (kg). A heavy object contains more matter than a lighter object. Think about an empty cardboard box – it is quite light because it doesn't contain much matter. If you fill the box with heavy items, it becomes much heavier. Obviously, as the box gets heavier, it becomes more difficult to move.

In other words, the heavier an object is, the more force you need to apply to get it to accelerate.

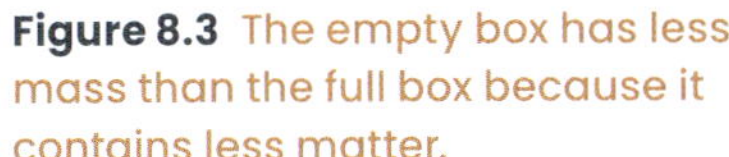

What is mass?

Figure 8.3 The empty box has less mass than the full box because it contains less matter.

❸ Force is a push, a pull or a twist

A **force** is a push, a pull or a twist. Force is measured in **newtons** (N). The size of a force is important. A person pulling on a broken-down car with a rope won't generate a large pulling force, and the car will experience only a very small acceleration. If you attach the car to a tractor, the car will experience a much greater acceleration because the tractor can generate a much larger pulling force than a person can.

The amount of acceleration an object experiences depends on how much force is applied to the object. The greater the force, the greater the acceleration.

What is the relationship between force and acceleration?

Figure 8.4 A tractor can apply a large force and so is able to accelerate a stationary car easily.

INVESTIGATION 8.1

Acceleration changes due to mass

KEY SKILL

Representing data to identify patterns and trends

▶ *Go to page* ***176***

CHECKPOINT 8.1

1 What is the formula used to calculate force?

2 Which box would have the greatest acceleration? Explain why.
  a Box A and box B have identical masses, but box B is experiencing a smaller force.
  b Box C and box D are both experiencing the same force, but box C is heavier.

3 A motorcycle is travelling at a constant speed. Is the motorcycle accelerating? Explain your answer.

4 Which has a greater mass – a bowling ball or a beach ball? Explain how you can tell.

5 Describe the relationship between acceleration, mass and force.

RESEARCH

6 A rocket is very heavy and undergoes a large acceleration as it takes off. So, the force on a rocket must be extremely large. Use the internet to find out how big the force is, and where this force comes from.

SUCCESS CRITERIA

- I can describe what is meant by acceleration, force and mass.
- I can predict how changing force and mass will change the acceleration of an object.

8.2 SPEED, DISTANCE AND TIME

LEARNING INTENTION

At the end of this lesson I will be able to explain the relationship between distance, speed and time.

KEY TERMS

distance
the total length an object travels

speed
the distance an object travels divided by the time taken to travel that distance

LITERACY LINK

SPEAKING

Write a script for a video that explains the relationship between speed, distance and time. With a partner, record your video.

NUMERACY LINK

UNITS

Convert 60 km to m, and 1 hour to seconds.

Then convert 60 kmh^{-1} to ms^{-1}. Show your working.

Convert 8 m to km, and 1 second to hours.

Then convert 8 ms^{-1} to kmh^{-1}. Show your working.

Figure 8.5 A fighter jet can travel large distances in very little time.

The fastest thing in the universe is light. Light can travel amazingly quickly – it can travel around Earth seven times per second! The speed of light in a vacuum is 300 000 000 ms^{-1}. This means that, in one second, light can travel 300 000 000 metres. This is extraordinarily fast, and it is very difficult to picture just how quick this is.

However, you can see from this that there is a clear relationship between **speed**, time and **distance**.

1 Speed is distance divided by time

Speed is the distance an object travels divided by the time it takes to travel that distance. Speed is often measured in metres per second or kilometres per hour. You can rewrite the definition of speed as an equation:

$$\text{speed} = \frac{\text{distance}}{\text{time}}$$

So, if a car travels 60 kilometres in 1 hour, its speed would be equal to 60 kmh^{-1}. From this equation, you can see that if object A travels a certain distance in less time than object B, then object A has a higher speed than object B.

What is speed?

❷ A formula triangle allows you to calculate speed, distance and time

You can calculate speed from the formula: speed = distance ÷ time. Speed is usually measured in metres per second (ms^{-1}), distance is usually measured in metres (m) and time is usually measured in seconds (s). To calculate distance or time, you can use a formula triangle (Figure 8.6).

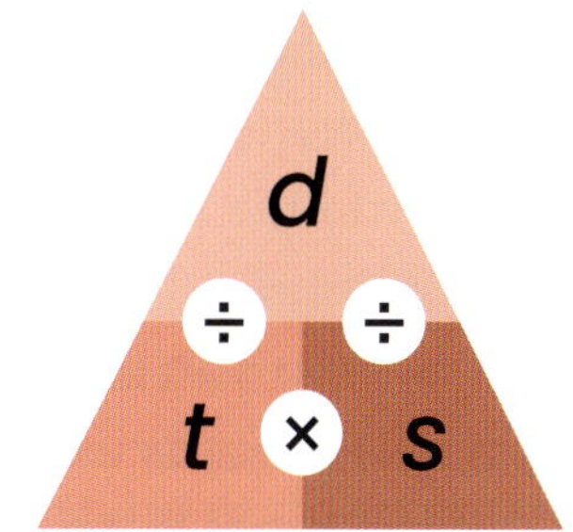

Figure 8.6 A formula triangle for distance (*d*), speed (*s*) and time (*t*)

To use the formula triangle, cover up the quantity you wish to find. What you can then see is the formula you need.

For example, if you know speed and time and wish to find distance, cover up the *d* on the triangle. This leaves $t \times s$. So, the formula is distance = time × speed. If you wish to find time, cover up the *t*, which leaves $d \div s$. So, the formula is time = distance ÷ speed.

How can you use a formula triangle?

❸ Distance is how far you travel

Imagine that one person is jogging and another person is sprinting. Assuming they maintain their speed for the entire time, who do you think will travel further in 1 minute? The sprinter will travel further because they will have been moving at a greater speed. Remember, speed is just a measure of how far you travel in a set amount of time – if you have a greater speed, you will travel further in that time.

To practise calculating distance, imagine you are riding a bicycle at a speed of 4 metres per second. How far would you travel in 3 seconds? Check the formula triangle – you can see that the formula for distance is $d = t \times s$. So, your distance travelled will be 3 s × 4 m/s = 12 m.

How is distance related to speed?

Figure 8.7 You can calculate the distance this cyclist travels if you know her speed and the time it takes her to travel that distance.

INVESTIGATION 8.2
Ticker timers

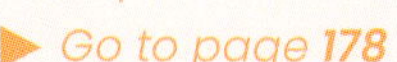

KEY SKILL
Identifying limitations to the method and suggesting improvements

▶ *Go to page 178*

CHECKPOINT 8.2

1 Describe how distance and time are related to speed.

2 What are the formulas for finding speed, time and distance?

3 Usain Bolt can run 100 metres in 9.58 seconds. Calculate his average speed across the 100 metres.

4 How long would it take to walk from Melbourne to Brisbane (a distance of 1600 km) at an average speed of 5 kmh^{-1}?

5 Sound can travel at 330 ms^{-1}. A spectator hears a starter's pistol 1.8 seconds after it was shot. How far away is the spectator from the pistol?

6 If you were riding on a roller coaster track, how would your speed change when you went:
a up a hill?
b down a hill?

7 Create a formula triangle for the equation from lesson 8.1: force = mass × acceleration

INQUIRY

8 Design an experiment that would allow you to measure the speed of light as accurately as possible.

SUCCESS CRITERIA

- I can explain the relationship between distance, speed and time.
- I can state the formulas to calculate distance, speed and time.

8.3 GRAPHING MOTION

LEARNING INTENTION

At the end of this lesson I will be able to describe the relationships between displacement, distance, time, speed, velocity and acceleration using visual representations.

KEY TERMS

displacement
the distance an object is from its starting position

scalar
a quantity that has a magnitude but no direction

vector
a quantity that has both a magnitude and a direction

velocity
a measure of how quickly displacement changes

LITERACY LINK

VOCABULARY

Create flashcards for the key terms listed above by writing the term on one side of a small card and the definition on the other. Study them for five minutes and then test a friend using the cards.

NUMERACY LINK

MEASUREMENT

Use a stopwatch to time how long it takes a pen to fall from your desk to the floor. Calculate the average speed of the falling pen.

The world's fastest person is Jamaican sprinter Usain Bolt. Bolt is nearly 2 metres tall and holds world records in the 100-metres sprint, 200-metres sprint and 4 × 100-metres relay. You can use the distance of his races, as well as the time he ran them in, to calculate his speed. You can also figure out his velocity by using his **displacement** and time.

Figure 8.8 Usain Bolt can reach a speed of 44.72 km per hour.

1 Distance is different from displacement

When discussing motion, it's important to use the right terms. Distance is how far an object actually travels, whereas displacement is how far the object is from its initial position.

Consider this scenario. From a starting position, you walk 6 metres north, then 8 metres west. This means that you have travelled a total distance of 6 + 8 = 14 metres. However, you are only 10 metres from your starting position, so your displacement is 10 metres northwest. Figure 8.9 shows another example of distance and displacement.

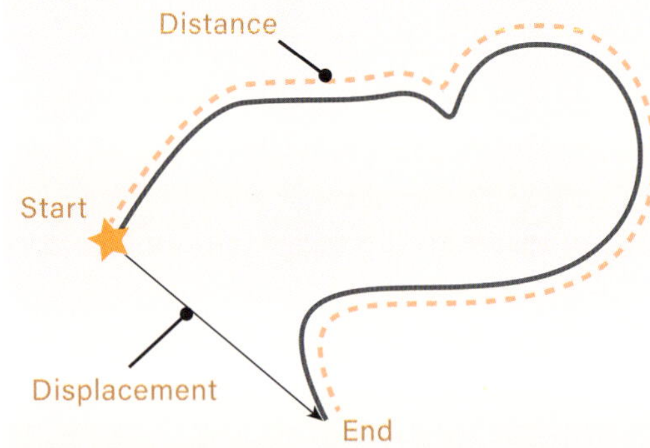

Figure 8.9 Distance and displacement can be very different.

Speed and velocity are both related to distance and displacement. Speed is a measure of how much distance changes in a certain time, while the **velocity** is a measure of how much displacement changes in a certain time.

In our example above, let's suppose you were walking for 10 seconds. Your average speed would be the distance travelled divided by the time taken, so 14/10 = 1.4 ms^{-1}. Your average velocity would be your displacement divided by the time taken, so 10/10 = 1.0 ms^{-1} northwest.

Displacement and velocity should always be given a direction as well as a magnitude (such as 6 ms^{-1} east). Quantities that have both a direction and a magnitude are called **vector** quantities. Other vector quantities include acceleration and force. Distance and speed do not require a direction, and are known as **scalar** quantities. Other scalar quantities include mass and energy.

❷ Distance–time graphs represent motion

Consider a student walking to class. In the first 10 seconds, she walks 15 metres. She then realises she's running late and jogs for the next 10 seconds, travelling 30 metres. Finally, she sprints the last 25 metres to class in 5 seconds. A distance–time graph of the student's journey would look like Figure 8.10. The average speed can be calculated by finding the gradient of the graph. For example, in the 0–10 second interval, the speed is 15/10 = 1.5 ms^{-1}. At each stage of this particular graph, the gradient increases. This tells us that the student is moving at a greater speed in each successive interval.

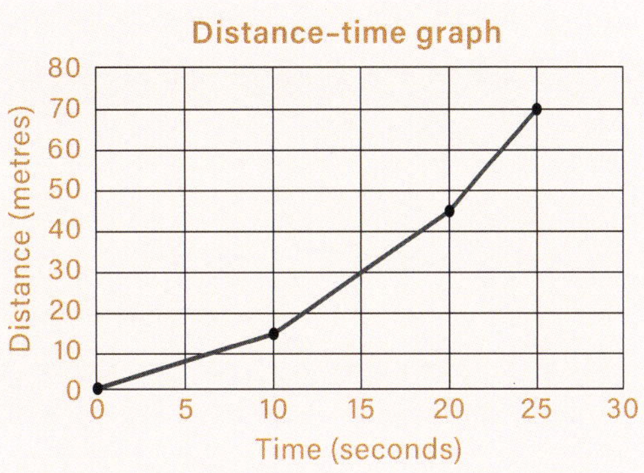

Figure 8.10 A distance–time graph of the student's journey

What does the gradient of a distance–time graph show?

❸ Speed–time graphs give more information

In the first interval, the student is moving at 1.5 ms^{-1}; from 10 seconds to 20 seconds, she's travelling at 30/10 = 3.0 ms^{-1}; and from 20 seconds to 25 seconds, she's travelling at 25/5 = 5 ms^{-1}. We can graph the speed she's travelling against the time, to create a speed–time graph.

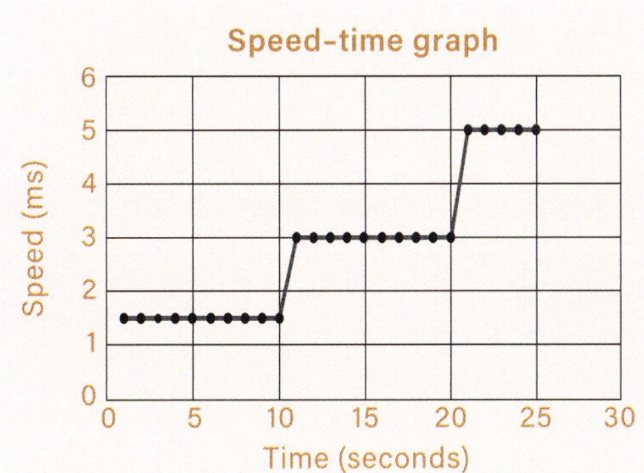

Figure 8.11 A speed–time graph of the student's journey

Between 0 and 10 seconds, the graph is flat, showing that the student was travelling at a constant speed of 1.5 ms^{-1}. Between 10 and 11 seconds, the graph increases, showing that the student is accelerating. In fact, the magnitude of the acceleration is given by the gradient of the line. In this case, the acceleration of the student is 1.5/1 = 1.5 ms^{-2}. Finally, the area of the underside of the graph will show the distance travelled. The area under the graph for the first interval in Figure 8.11 is 1.5 × 10 = 15 m, which is exactly how far the student travelled in Figure 8.10.

What does the gradient of a speed–time graph show?

INVESTIGATION 8.3

Vehicles and pedestrians

KEY SKILL
Representing and recording data using a table

▶ *Go to page* ***180***

CHECKPOINT 8.3

1 Explain the difference between distance and displacement using an example.

2 What is velocity and how is it different to speed?

3 What does a decreasing gradient on a speed–time graph demonstrate?

4 a A person walks 8 metres north and then 2 metres south. Give both the distance and displacement for this scenario.
  b If it took this person 8 seconds to complete the walk, calculate both their speed and velocity.

EXTENSION

5 Usain Bolt is the world's fastest man. Research his incredible 200-metre world record race in 2009. Calculate Bolt's average speed and velocity during this race, using a diagram of the track as well as the calculations for speed and velocity. Hint: You will need to find his displacement first.

SUCCESS CRITERIA

- I can explain the difference between distance and displacement and provide an example diagram of each.
- I can explain both speed and velocity and give the formulas for both.
- I can interpret speed–time and distance–time graphs.

8.4 NET FORCE AND ACCELERATION

LEARNING INTENTION

At the end of this lesson I will be able to relate acceleration to a change in speed and/or direction as a result of a net force.

KEY TERMS

friction
a contact force that opposes motion, caused by objects rubbing against each other

net force
all the forces acting on an object added together

normal force
the force of the floor or ground pushing an object upwards

weight force
the force generated by gravity pulling an object downwards

NUMERACY LINK

GRAPHING

Andy records the force applied to a toy car and its acceleration.

Force (N)	Acceleration (ms^{-2})
0.105	0.5
0.21	1
0.315	1.5
0.42	2

Draw a line graph to display this data. What does the gradient tell us?

Think about the forces acting on you right now. The first one you will probably think of is gravity, the force that pulls you down to the ground.

Recall that the acceleration of an object is related to the force acting on the object as well as the mass of the object. So, if there is force acting on you, and you have a mass, why don't you accelerate? The answer is in a concept called 'net force'.

1 Net force is the sum of all forces acting on an object

The **net force** is all the forces acting on an object added together. Consider the forces acting on a cyclist. The driving force of their feet on the pedals is propelling them forward. Gravity is also acting on them, providing a **weight force**. **Friction** between the tyres and the road opposes their motion, as well as air resistance. Finally, there is the force from the road pushing the bicycle upwards, which is called the **normal force**.

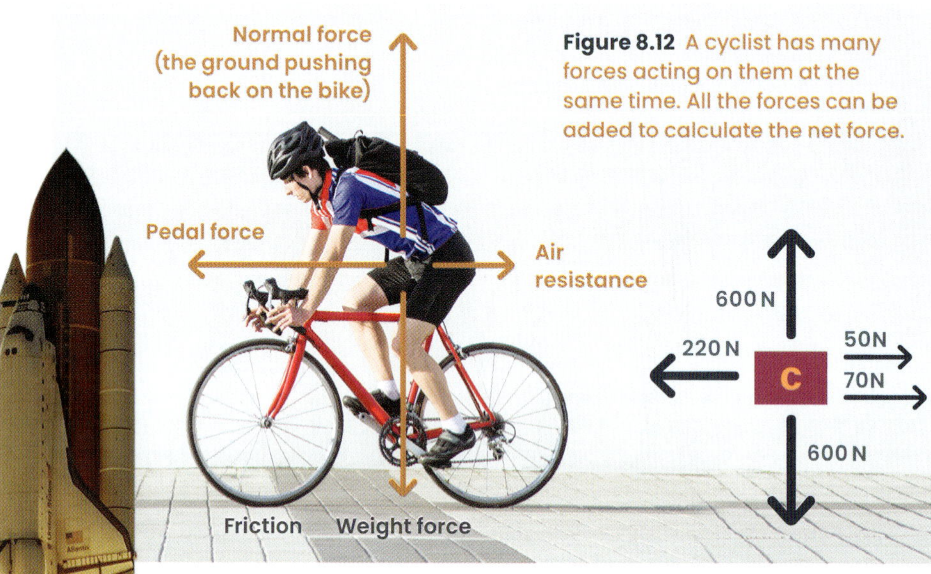

Figure 8.12 A cyclist has many forces acting on them at the same time. All the forces can be added to calculate the net force.

Figure 8.13 The large mass of the rocket means it requires a huge force to increase its speed.

To work out the net force acting on the cyclist, add together all the forces acting in the same direction and subtract forces acting in the opposite direction.

In the vertical direction, the weight force (600N) is acting in the opposite direction to the normal force (600 N), so subtract these: 600 N – 600 N = 0 N. In the vertical direction, the sum of the forces is zero, so you can ignore them. Air resistance (50 N) and friction (70 N) act in the same direction, so add these together. The total force acting to the right is 50 N + 70 N = 120 N. The pedal force is to the left, which is the opposite direction to the friction and air resistance, so subtract them. Therefore, the net force is 220 N – 120 N = 100 N to the left.

What is a 'net force'?

2 Acceleration is any change in speed or direction over time

People often mistakenly think of acceleration as meaning 'speeding up'. However, acceleration is any change in speed or direction over time – speeding up, slowing down or turning.

If acceleration is in the same direction as an object's motion, the object will speed up. If acceleration is in the opposite direction to an object's motion, the object will slow down. If acceleration is not parallel to the object's motion, the object will change direction. Acceleration is usually measured in metres per second squared (ms^{-2}).

Does acceleration result in a faster speed?

3 Acceleration and net force always act in the same direction

When many forces are acting on an object, the acceleration depends on the net force and is always in the same direction as the net force.

If there is no net force, there will be no acceleration. If you are sitting in a chair reading this right now, you are not accelerating and so you have no net force acting on you. You have a weight force acting downwards and an equal-sized normal force acting upwards, but these two forces are in opposite directions so they cancel out.

What is the net force acting on a stationary object?

Figure 8.14 Net force and acceleration always act in the same direction.

INVESTIGATION 8.4
Balloon rockets

KEY SKILL
Identifying and managing relevant risks

▶ *Go to page **181***

CHECKPOINT 8.4

1 Define net force.

2 An object is accelerating downwards. In which direction must the net force on it be acting? Give evidence to support your answer.

3 Draw a diagram and calculate the net force when an object experiences:
  a forces of 20 N to the left and 30 N to the right
  b a driving force of 15 N to the left and forces of friction of 10 N and air resistance of 5 N to the right
  c a driving force of 20 N to the left, a normal force of 8 N upwards, a weight force of 8 N downwards, and forces of friction of 10 N and air resistance of 5 N to the right.

4 A jet plane is travelling at a constant speed of 800 kmh^{-1}.
  a Is the plane accelerating?
  b Given your answer to part a, what must the net force acting on the plane be?

STUDENT VOICE AND AGENCY

5 Prove that acceleration does not always cause an object to speed up. You can do this using any method you like; for example, creating a model, a poster, an experiment or a podcast. Negotiate with your teacher how your work will be assessed.

SUCCESS CRITERIA

- I can describe how an object will accelerate, given the direction of its net force.

8.5 NEWTON'S LAWS

LEARNING INTENTION

At the end of this lesson I will be able to qualitatively analyse everyday situations involving motion in terms of Newton's laws.

KEY TERMS

action–reaction pair
two equal and opposite forces exerted by two objects on each other

balanced force
a force acting on an object that is cancelled out by another force, so the net force is zero

inertia
a property of matter that causes it to resist change in speed or direction (to remain at rest or in a state of uniform motion)

stationary
not moving; still

unbalanced force
a force acting on an object that is not cancelled out by another force, so a net force acts on the object

LITERACY LINK

WRITING

Design a poster that summarises Newton's three laws of motion.

NUMERACY LINK

CALCULATION

Calculate the force required by the propeller of a small 520 kg plane if it accelerates at 7 ms^{-2} during take off.

Isaac Newton was an English scientist who lived in the 1600s. He is so well regarded that the unit of force is named after him. Among his many achievements are his three laws of motion, which describe how objects move in relation to the forces applied to them.

Understanding Newton's laws helps you predict what will happen to objects experiencing forces in any situation.

Figure 8.15 Newton's third law explains that while Earth pulls a skydiver down, the skydiver also pulls Earth up!

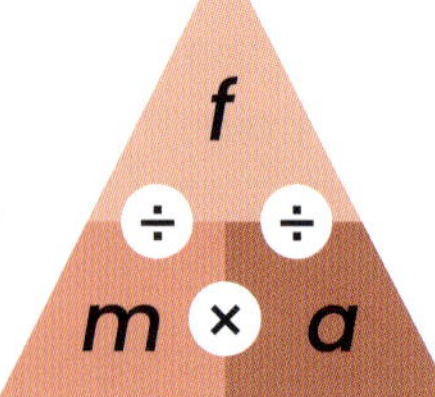

Figure 8.16 A formula triangle for force (f), mass (m) and acceleration (a)

1 Newton's first law: an object stays at rest or at the same speed unless an unbalanced force acts on it

Newton's first law states that an object will continue doing what it is doing unless acted on by an **unbalanced force** – a force that causes a change in motion.

This means that an object that is **stationary** (not moving) will stay stationary, and a moving object will keep moving unless a force acts on it. You rarely see this happening – if you roll a ball across the floor, it doesn't keep rolling forever. Rather, it slows down and eventually stops. This is because there is an unbalanced force acting on the ball – friction. The ball is being slowed down by a combination of friction from the ground and air resistance. If you rolled the ball where there is no friction (such as in space), it would not stop until it was acted upon by another force. The property of an object to keep doing what it is doing is called **inertia**.

A car travelling at a constant speed also experiences friction. In this case, the friction from the road is a **balanced force** – it is exactly equal to the thrust from the engine – so the car does not slow down. The backwards force on the car is exactly the same size as the forwards force, and so the forces are balanced.

What is Newton's first law?

Figure 8.17 If you threw an object in deep space, it would keep going, possibly for hundreds of thousands of years, until other forces acted upon it.

❷ Newton's second law: force = mass × acceleration

Newton's second law states that the size of the acceleration of an object is proportional to the force applied and inversely proportional to the mass of the object. In other words, the heavier an object is, the more force is required to accelerate it.

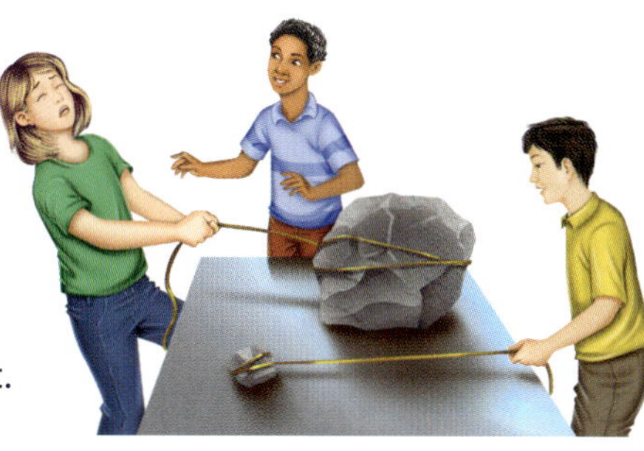

Figure 8.18 The heavier an object is, the more force is required to accelerate it.

Imagine you are standing on a skateboard and you use your foot to propel yourself. The force of your foot against the ground causes your acceleration. If someone else is standing on the skateboard with you, the total mass increases, so if you apply the same force with your foot, you will not accelerate as much. If you wanted to accelerate at the same rate as before, you would need to apply much more force.

What is Newton's second law?

❸ Newton's third law: for every action, there is an equal and opposite reaction

Newton's third law is possibly the most famous. It says that for every action, there is an equal and opposite reaction. This means that if object A exerts a force on object B, object B exerts the same sized force on object A, but in the opposite direction.

When you jump, you push the ground down with your legs, but the ground also pushes you up – that's why you can jump. This is an example of Newton's third law – the action (you pushing the ground down) has an equal and opposite reaction (the ground pushing you up).

Every force has an equal and opposite reaction force. The tyres on a car push the road backwards, and are in turn pushed forwards by the road. You are pushing down on your seat right now, and your seat is pushing you up, preventing you from falling onto the ground. When you jump, Earth's gravity pulls you back down, but you also pull Earth up very slightly. The equal and opposite forces in these examples are known as **action–reaction pairs**.

Figure 8.19 As you push down on Earth, Earth pushes up on you.

What is Newton's third law?

INVESTIGATION 8.5
Car safety

KEY SKILL
Identifying the variables and formulating a hypothesis

▶ *Go to page **182***

CHECKPOINT 8.5

1 Identify whether these definitions apply to Newton's first, second or third law.
  a The acceleration of an object depends upon the force applied and the mass of the object.
  b For every action, there is an equal and opposite reaction.
  c An object keeps doing what it's doing unless acted on by an unbalanced force.

2 Newton's first law states that an object keeps doing what it's doing unless acted on by an unbalanced force. So, why must you continue to pedal a bicycle if you want to remain at the same speed?

3 What is the equal and opposite force acting when a:
  a cricket bat pushes a ball forwards?
  b person pulls a suitcase?

4 If you push against a wall with 45 N of force, and neither you nor the wall moves, how much force is the wall pushing you with?

RESEARCH

5 Isaac Newton was famous for more than his laws of motion. Use the internet to research more of Newton's scientific achievements.

SUCCESS CRITERIA

- I can state each of Newton's three laws.
- I can use Newton's three laws to explain the motion of objects.

VISUAL SUMMARY

Newton's first law:
An object will continue in the same state of motion unless an unbalanced force acts upon it.
- If an object is not moving, it will stay stationary.
- If an object is moving at a constant speed, it will continue to move at that same speed.

A stationary object will stay stationary as long as no unbalanced forces act on it.

A moving object will keep moving as long as no unbalanced forces act on it.

Sir Isaac Newton ▲

Newton's second law:
The acceleration of an object increases with the force applied, and decreases with the mass of the object.

Force = mass × acceleration

More force = more acceleration

Newton's third law:
Every action has an equal and opposite reaction.

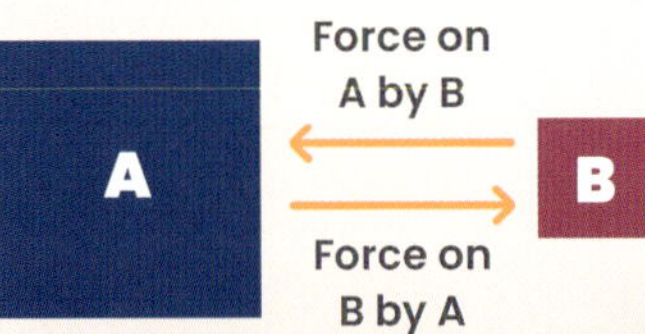

✓ **Distance** is how far an object actually travels, while **displacement** is how far the object is from its initial position.

✓ **Speed** is how much distance changes in a certain time, while **velocity** is how much displacement changes in a certain time.

For an object to accelerate, there must be an unbalanced force. The object will always accelerate in the direction of the unbalanced force.

Car speeds up

Direction of motion

Force

Car slows down

Direction of motion

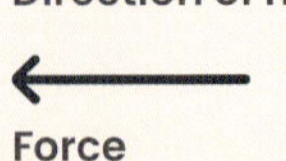

Force

Car turns

Direction of motion

Force

The speed of an object is equal to the distance travelled divided by the time taken.

Speed = distance ÷ time

To calculate distance, speed or time, you can use a formula triangle.

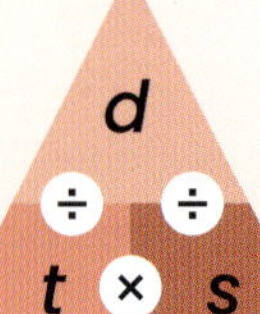

★ FINAL CHALLENGE ★

1. Copy and complete the following sentences.
   - a An object will continue doing what it's doing unless it is acted on by an unbalanced ______________.
   - b The acceleration of an object increases with a larger ______________ and decreases with a larger ______________.
   - c For every ____________, there is an equal and opposite ____________.

2. Suggest how you can increase the acceleration of an object.
3. What is the formula to calculate each of these? (Hint: Use your formula triangle.)
   - a speed
   - b time
   - c distance

4. What is the net force acting on objects A, B and C?

A: 4 N ←, 6 N →
B: 4 N ↑, 2 N ↓, 9 N ←, 3 N →, 6 N →
C: 5 N ↑, 5 N ↓, 4 N ←, 2 N ←, 3 N →, 3 N →

5. For each object, A, B and C, specify the direction of acceleration.
6. a A train completes a trip of 240 km in 2.5 hours. What was the average speed of the train?
   - b The train driver thinks that he can increase the speed of the train to 140 kmh^{-1}. How long would the journey take if the train travelled at this speed?

Level 4

200xp

LEVEL UP!

7. Match each of Newton's laws with the scenario that demonstrates it.

Newton's first law	A fully loaded car takes longer to get up to top speed.
Newton's second law	As your feet push down on a trampoline, you jump higher.
Newton's third law	If you quickly pull a sheet of paper out from under a coin, the coin will remain where it is.

8. Using your understanding of Newton's first law, explain why a rocket requires much more fuel to lift off than it does to change direction once it is in space.
9. Suggest a way in which each of Newton's three laws has affected something you have done today.

SKILLS AND INVESTIGATIONS

SCIENCE SKILLS

Questioning, predicting and planning

Collecting and using data

Analysing and visualising data

Uncertainty, errors and mistakes

Science in the media

Writing investigation reports

INVESTIGATIONS

BIOLOGICAL SCIENCES

1.1 Genetic trait survey

1.2A Extracting DNA from strawberries

1.2B DNA lolly model

2.2 Natural selection in practice

2.3 Environmental and genetic factors affecting survival

2.4 Comparative anatomy

2.5 Dating fossils

CHEMICAL SCIENCES

3.4A Physical properties of elements

3.4B Chemical properties of elements

4.5A The effect of concentration on the rate of a reaction

4.5B The effect of temperature on the rate of a reaction

4.6A The effect of a catalyst on the rate of a reaction

4.6B The effect of surface area on the rate of a reaction

EARTH AND SPACE SCIENCES

5.1 Observing interactions between the spheres

5.2A Releasing dinosaur breath

5.2B The effect of temperature on soil respiration

5.3 Investigating sunscreens

5.4 The greenhouse effect

6.1 Modelling the expanding universe

6.2 Stargazing

6.4 Investigating orbits

6.5A Making a telescope

6.5B Lens diameter and resolution

PHYSICAL SCIENCES

7.1 Galileo's pendulum

7.2 Energy efficiency of bouncing balls

7.4 Insulation and heat transfer

7.5 Atmospheric convection and the inversion layer (Teacher demonstration)

8.1 Acceleration changes due to mass

8.2 Ticker timers

8.3 Vehicles and pedestrians

8.4 Balloon rockets

8.5 Car safety

KEY SKILLS

Skill	Investigations
Using primary and secondary scientific evidence	Investigation 1.1
Identifying and managing relevant risks	Investigation 1.2A Investigation 4.5A Investigation 8.4
Using modelling and simulations	Investigation 1.2B Investigation 2.5 Investigation 6.1
Representing data to identify patterns and trends	Investigation 2.2 Investigation 4.5B Investigation 5.4 Investigation 6.4 Investigation 6.5B Investigation 8.1
Identifying the variables and formulating a hypothesis	Investigation 2.3 Investigation 4.6B Investigation 7.2 Investigation 8.5
Explaining results using scientific knowledge	Investigation 2.4 Investigation 3.4A Investigation 5.2A
Writing a research question	Investigation 3.4B Investigation 5.1 Investigation 6.5A
Drawing conclusions consistent with evidence	Investigation 4.6A
Identifying the independent, dependent and controlled variables	Investigation 5.2B
Evaluating results for reliability and validity	Investigation 5.3
Referencing sources of information	Investigation 6.2
Representing and recording data using a table	Investigation 7.1 Investigation 8.3
Identifying limitations to the method and suggesting improvements	Investigation 7.4 Investigation 8.2

QUESTIONING, PREDICTING AND PLANNING

KEY TERMS

controlled variables
all the things that need to stay the same during an investigation

dependent variable
the thing that will be measured and is altered by the independent variable

experiment
an investigation carried out under controlled conditions, to test a hypothesis

fair test
an investigation in which only one factor is changed and all other variables are kept the same

fieldwork
an investigation conducted in the natural environment, not a laboratory

hypothesis
a scientific statement that can be tested

independent variable
the thing that is purposefully changed during an investigation

reliable
provides consistent results when repeated

research
to gather data and information in an organised way to inform a hypothesis or an investigation

valid
measures what is intended to be measured

Science is all about investigating – asking questions, looking at data and drawing conclusions about how things work. A scientist is like a detective, but instead of investigating a crime, they're investigating the world. To be useful, a good scientific test needs to follow certain principles.

1 Good science needs to be valid and reliable

When scientists design investigations, they ask themselves 'is this *good science*?'

To figure out if something is good science, you need to check that it's both **valid** and **reliable**. If a test is reliable, you can do the test over and over again and get very similar results. If a test is valid, it measures what it is supposed to measure.

Imagine you design a catapult that launches marshmallows and decide to test it against a friend's design. Just as your friend is firing the catapult, a massive gust of wind blows their marshmallow further than yours – that's not fair, right? It's not a valid outcome because the wind caused the increased distance, not the catapult. The test didn't measure what you wanted it to measure (the power of the catapult); it measured the power of the catapult *and* the power of the wind. It's not reliable because, if you did the test again, the wind might be weaker, stronger or not there at all.

Why does good science need to be valid and reliable?

2 A fair test needs to be controlled

Fair tests are essential for good science. A **fair test** is one in which only one variable is changed and all other variables are kept the same. Variables are the things that can be controlled, changed or measured during an investigation or experiment. There are three main types of variables: independent, dependent and controlled variables.

The **independent variable** is the one thing you want to change in an investigation. If you change more than one thing, the investigation probably won't be a fair test. The **dependent variable** is what you are measuring in an investigation, and is what is altered by the independent variables. Examples include time in seconds or mass in grams.

Experiments are carried out in order to test a hypothesis.

The **controlled variables** are all the things you will keep the same. Examples of controlled variables are temperature, mass, equipment, location and volume.

Let's say you decide to put three plants in three different amounts of sunlight to see which plant grows the most. You would make sure the plants were the same size, health and species, and only change the amount of sunlight the plant is getting – this is the independent variable. The dependent variable would be your measurement of the plants' growth (which could be their weight or their size) and the controlled variables are all the other factors.

What are the three types of scientific variable?

3 A hypothesis is a prediction of the outcome

A **hypothesis** is a prediction made to test something. A good hypothesis involves some reading and research so that scientists can make an informed decision about what they think will happen, before testing it in an investigation. A hypothesis can be supported (found to be correct) or rejected (found to be incorrect).

You use the independent and dependent variables when writing a hypothesis, so the first step is always to identify these. The general rule to use when writing a hypothesis is:

Independent variable

If I do this, then this will happen.

Dependent variable

Even though this rule has the word 'I' in it, that's not how you write the hypothesis! You should always write it formally and in the third person (don't use *I*, *we*, *you* etc.).

What is a hypothesis?

These students are doing fieldwork to test water samples.

ELEMENTS OF AN INVESTIGATION

Think again about the investigation that involves plants in different amounts of sunlight to see which plant grows the most. The elements of this investigation are:

- *hypothesis*: If a plant is placed in direct sunlight, then it will grow more than a plant in indirect or no light.
- *independent variable*: amount of direct sunlight (one plant is put in a dark cupboard, one is put outside in direct sunlight and one is put near a window)
- *dependent variable*: growth of the plant, in millimetres
- *controlled variables*: species of plant, starting size of plant, health of plant, amount of water given to plant.

TYPES OF INVESTIGATIONS

Scientists do many different types of investigations, depending on their area of science and the information they need to gather.

Fieldwork happens when information and data are collected outside of the laboratory or usual setting. Environmental scientists often do fieldwork, such as collecting water samples from streams to study the water quality or counting the numbers of species of plants and animals in an area.

Experiments are usually carried out to test a hypothesis. Experiments in science include (among other things) those undertaken in chemistry, physics, earth science, and with living things in biology.

Research informs a hypothesis before it is created. Scientists often share their research so they can build scientific understanding and discoveries over time.

COLLECTING AND USING DATA

Data is like evidence – you need it to draw your conclusion. Scientists collect and analyse data to test their hypotheses.

KEY TERMS

accuracy
how close the results are to the true values

inference
an educated guess or judgement based on observations

observation
something you see and know to be true

prediction
a statement about the future based on observation and evidence

primary data
first-hand data, from your own investigation

qualitative
written descriptions and observations

quantitative
numerical information and data

sample size
the number of participants in a survey or samples tested in an experiment

secondary data
second-hand data, from someone else

CALCULATING THE MEAN

To calculate the mean (also known as the average) of a group of numbers, add all the numbers together and then divide them by how many numbers you added together.

For example, to calculate the average time taken in Table 1, you would add 174, 110 and 48, then divide by 3.

The mean would be $(174 + 110 + 48) \div 3 = 110.7$ s

1 Scientists collect different types of data

One way to describe data is that it can be **qualitative** or **quantitative**. Quantitative data relates to quantities – that is, numbers. Quantitative data can include the number of something, the volume, the length, time, or anything that scientists can physically measure or count. Qualitative data relates to the qualities of something – that is, written descriptions or observations about data.

Another way to describe data is as **primary data** or **secondary data**. Primary data is first-hand data that you collect yourself through scientific investigation. Secondary data is second-hand data, gathered by someone else and given to or accessed by you.

To make sure secondary data is valid and reliable (see page 126), you need to check that it comes from a reliable source. It's also important to make sure the data is **accurate**. If it is a survey, was the **sample size** large enough or did it only involve a small number of people? Are the results from just one country or population group?

How are qualitative and quantitative data different?

2 Data needs to be carefully collected and recorded

To ensure that their data is valid, scientists record observations and measurements very carefully. They might do this in a logbook or table for quantitative data. Qualitative data might be recorded in a journal or workbook. Sometimes data is visual and can be recorded with a camera.

When taking measurements and recording quantitative data, use the appropriate units for physical quantities. This table shows some common metric units for physical quantities.

Physical quantity	Measurement and unit	Conversion
Length	Millimetre (mm)	10 mm = 1 cm
	Centimetre (cm)	100 cm = 1 m
	Metre (m)	1000 m = 1 km
	Kilometre (km)	
Mass	Milligram (mg)	1000 mg = 1 g
	Gram (g)	1000 g = 1 kg
	Kilogram (kg)	
Volume	Millilitre (mL)	1000 mL = 1 L
	Litre (L)	
Temperature	Celsius (°C)	

Why is it important to use the correct units when measuring?

❸ Organising data makes it easier to understand

Collecting data isn't the end of the process – the data needs to be analysed and considered. That means it must be well organised and clearly presented, or else it will be difficult to understand.

One of the best ways to arrange and present scientific data is in a table. Always design and rule out your table before you start your investigation – this ensures you are ready and organised to collect the correct data, and that you don't forget to collect important data.

MAKING A GOOD SCIENTIFIC DATA TABLE

1 Use a ruler so that your table is clear and easy to read.
2 Give your table a descriptive and useful title and include a table number, in case you want to refer to it in your investigation report.
3 Include the units in the column headings where needed (e.g. mm).

Clear descriptions in titles

Table has a number and title

TABLE 1: TIME TAKEN TO DISSOLVE AN ALKA-SELTZER TABLET

Type of water	Water temperature (°C)	Time taken (s)
Cold	4	174
Room temperature	20	110
Warm	40	48

Units are given at the top of each column

Lines are ruled and easy to follow

Another way to present data is to illustrate it using a chart or graph. This is an excellent visual way to show the data from your table.

Why are tables used to organise data?

Charts, graphs and other visual tools can be used to present and explain data.

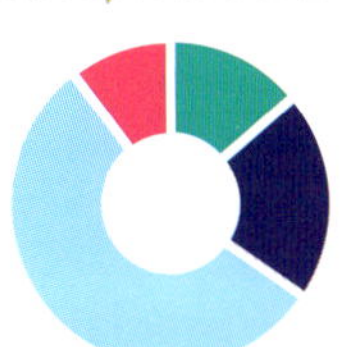

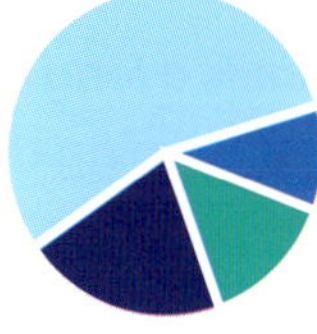

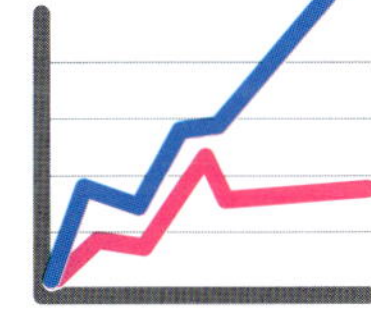

Recording your observations is a great way to gather first-hand data.

INFERENCES, OBSERVATIONS AND PREDICTIONS

An **inference** is something you think might be the case, but you don't know for sure. An **observation** is something you see and know is definitely true. A **prediction** is what you think will happen in the future.

In science, an observation can often lead to an inference. You could observe that your cactus is dying, and then infer that this was because it was overwatered. You could then stop watering it and observe it again – this could re-inform and change your inference.

ANALYSING AND VISUALISING DATA

KEY TERMS

mean
the values of a data set added together and divided by the number of values

median
the middle result in a set of data

precision
how close the results of repeated trials are to each other

range
the largest value in a data set minus the smallest value

trend
a consistent change in the values of a data set

TYPES OF DATA

Discrete data is data that can only have certain values. For example, the number of seeds that sprouted in a week. You can't have half a seed sprouting, so our data here can only have the value of whole numbers.

Continuous data is data that can have any value, including decimal points. For example, the time taken for a ball to roll a certain distance. The time could be in seconds, or tenths of seconds, or hundredths or thousandths of seconds. Any data that can take any possible value is continuous data.

After completing an experiment, you now most likely have a table showing the raw data that you collected. But a table full of numbers might not tell you much about what your results mean, so now it's time for some analysis.

1 Raw data needs to be analysed

Trial	Distance (cm)
1	235
2	244
3	242
4	238
5	133

There are many ways that data collected from an experiment can be analysed. Let's say you performed an experiment to measure how far a catapult will launch a ball. The results are shown in the table on the right. What can we tell from this data?

The **mean** is the average of a data set. To calculate the mean, simply add up the values and divide by the number of values. In the case of this data set, the mean is:

$$(235 + 244 + 242 + 238 + 133) \div 5 = 218.4 \text{ cm}$$

The mean is a useful way of determining a predicted value if the experiment were to be repeated.

The **median** is the middle result in a set of data. Using the median rather than the mean offsets the effect of extreme outliers. In the catapult results, the fifth trial clearly travelled a much shorter distance than the others. It appears that something may have gone wrong during this trial – perhaps a misalignment of the catapult or a strong wind affecting the distance. This means that the mean (218.4 cm) is smaller than four out of the five measurements – not ideal when talking about expected values! In this case, it may be better to use the median result. To determine the median, arrange the values in order from smallest to largest (133, 235, 238, 242, 244) and the middle value (238) is the median. 238 cm for this data set is probably much closer to the expected value than our mean of 218.4 cm would be. If there had been an even number of trials, then the median would be calculated by finding the mean of the two middle values.

The **range** of values is the difference between the largest and smallest value. In this example, our range is:

$$244 - 133 = 111 \text{ cm}$$

The range can tell you a lot about the **precision** of the experiment – the smaller the range, the more precise our experiment is. In this case, we have a very large range, and so a very low precision.

The fifth result in this data set is an outlier – a result that is a long way from the other results. If you notice an outlier while performing the experiment, the best approach would be to disregard that particular result and redo it. Hopefully, redoing the trial will give you a result closer to what was expected. If you can't redo the trial (for example, if you've already packed away your equipment before noticing the outlier), it can be either included (if it won't affect the outcome too badly) or excluded from your results. In this case, the outlier is so far from our other results that it makes sense to exclude it – it is clearly an incorrect value. If we removed the fifth trial, we would have a mean of 239.75 cm, a median of 240 cm (the mean of the two middle values) and a range of 6 cm. In short, our data would become much more precise and accurate. However, we cannot just pretend that the fifth trial didn't happen – someone reading your report might understandably wonder why you have five trials for most data points and only four for that point. If you choose to disregard a trial, you must address it in your discussion. You need to explain what the result was, and why you have chosen to remove it.

What is the difference between the mean and the median?

2 Visualising data can show patterns and relationships

In an experiment, we need to have an independent variable that we change and a dependent variable that we measure. In this example, we have performed an experiment where we changed the angle of the catapult (independent) and measured the distance a ball was thrown (dependent). A good experiment will use repeat trials for each value of the independent variable. The data table for this experiment might look like this:

	Distance travelled (cm)					
Angle (°)	**Trial 1**	**Trial 2**	**Trial 3**	**Trial 4**	**Trial 5**	**Mean**
5	135	142	138	145	140	140.0
10	195	188	193	188	192	191.2
15	235	244	242	238	133	218.4
20	287	295	290	289	286	289.4
25	340	339	344	339	335	339.4

It can be difficult to analyse data that is just raw numbers such as above. To help us see patterns, it is usually a good idea to display the data in a graph or chart. A line graph of the above data looks like that shown at the right.

This graph now makes it much easier to see the pattern (or **trend**) in our data – as the angle of release is increased, the distance travelled will increase as well. A strong analysis will then describe the relationship between the independent and dependent variables.

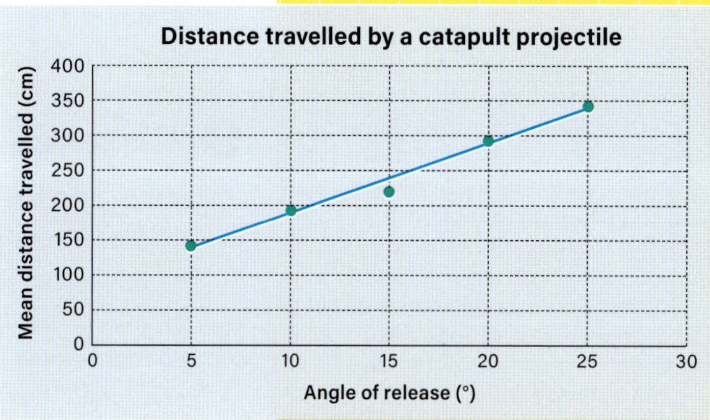

How do graphs help us to understand data?

DRAWING A LINE GRAPH

A scientific line graph should have:

- a title
- a consistent scale and axis label on both the *x*-axis and the *y*-axis: usually, the *x*-axis is the independent variable and the *y*-axis is the dependent variable. Make sure to include units in your axis labels
- a line of best fit: this is a straight line that shows the pattern of your results. Your line of best fit doesn't have to go through every data point, but should show the general direction of your data. Your line of best fit should never change direction or zig-zag – it should just be a single straight line.

UNCERTAINTY, ERRORS AND MISTAKES

KEY TERMS

error
the difference between the measured value and the actual value

mistake
something that has been done incorrectly in an experiment

uncertainty
the margin of error of a measurement

EXPERIMENTAL ERRORS

There are two categories of errors that could occur in an experiment.

Systematic errors are errors that will give consistent, incorrect results. Systematic errors can include things like incorrectly calibrated equipment or a flawed experimental design. These errors can be difficult to detect, as repeated measurements may yield the same results. Experiments must be carefully designed to avoid systematic errors. Comparing your results to expected results may also help identify systematic errors.

Random errors give inconsistent results, and repeated measurements will give different values. Examples of random errors include human reflexes when using a stopwatch or fluctuating values on a measuring instrument. The easiest way to reduce the impact of random errors is to take repeated measurements and then average them out.

Three things that can lead to problems in a set of data are **uncertainty**, **errors** and **mistakes**. While they all sound very similar, in a scientific experiment, these three terms have important different meanings.

1 Uncertainty depends on the measuring device

Uncertainty is the limit to how accurate a measurement can be due to the measuring device. The scale on the right shows that the chocolate weighs 100 g. However, as the scale only measures to the nearest whole number of grams, it doesn't tell us the whole story. The chocolate could weigh anywhere from 99.5 g to 100.4 g and the scale will round it to 100 g. There are a range of possible values of the mass of the chocolate and so there is uncertainty about its true mass.

In a scientific report, we would say the value of the mass is 100 ± 0.5 g. That is, the true value is within 0.5 g of 100 g. We call this ± 0.5 the uncertainty of the measurement. What if the scale showed 100.0 g? The chocolate would still have a possible range of values from 99.95 g to 100.04 g, and our recorded value would be 100.0 ± 0.05 g. In general, the uncertainty of a measurement is ± half of the smallest reading on the measuring device. If your device measures to the nearest gram, then the uncertainty is ± half a gram.

What determines the uncertainty of a measurement?

2 Errors can be systematic or random

Experimental error is how far away your measured value is from the actual value. There are two categories of errors: systematic errors, which give consistently incorrect results, and random errors, which give inconsistently incorrect results.

What is the difference between a systematic error and a random error?

3 Mistakes are caused by humans

Mistakes are things that you have done incorrectly in your experiment; for example, reading a thermometer incorrectly or adding too much of a certain chemical. Mistakes may invalidate your results, depending on how severe they are. Mistakes should not be written up as a part of your error analysis, and instead the experiment (or the part of it that contained the mistake) should be redone and the mistake avoided.

What is the difference between a mistake and an error?

SCIENCE IN THE MEDIA

Today in the digital era, there is an unimaginably large amount of scientific information available to be interpreted (correctly or not!) by anyone with internet access. How can we tell what is true or false, or, in more scientific terms, what is valid or invalid? How does the general public interpret science in the media and how can we ensure we critically evaluate all that we read and watch online?

1 Media reporting of science is not always reliable

When considering whether a science report in the media is trustworthy, asking yourself some questions can help.

- Is the headline **sensationalised**? Is the author just trying to get people's attention no matter what?
- Have the results been misinterpreted – possibly by someone without scientific understanding?
- Is the sample size small? (This could lead to invalid interpretations of data.)
- Does the article have **speculative** language – or is it written with the facts only?
- Are reputable scientists quoted in the article?

If the article references other information or research, you can follow up on its sources to see if they are trustworthy, valid and reliable.

Why is trustworthy science reporting important?

News media is an important method for science to be communicated to the public, but sometimes the science is overtaken by the story, leading to sensationalised headlines that can mislead the reader.

KEY TERMS

sensationalised
language that aims to shock or produce an emotional reaction in the reader

speculative
relying on theories or opinions rather than facts

ETHICAL CONDUCT

In general terms, ethics is the consideration of what is right or wrong so that the best possible choice can be made. Scientists can further ensure ethical conduct by reporting all scientific research honestly and accurately, by following safety standards and by ensuring their investigations are valid and reliable – acting to reduce and eliminate errors. If any activity includes experimental investigations using human or animal subjects, a legal and moral responsibility exists to follow ethical principles. If animals are used, further research to ensure that all legislation is being followed is important. Students can read more in the *Prevention of Cruelty to Animals Act 1986* and its *Regulations 2019* as well as the *Australian Code of Practice for the Care and Use of Animals for Scientific Purposes 2014*.

WRITING INVESTIGATION REPORTS

Writing an investigation report is a key skill in science, and one you will use many times during scientific study. By writing a clear, consistent report at the end of your investigation, you ensure that other people will understand your work.

1 An investigation report has a consistent structure

Your investigation reports should typically have a similar structure to the one shown here.

The title should be clear and in plain language. Many scientists write their title as a research question.

HOW DOES TEMPERATURE AFFECT THE RATE OF A CHEMICAL REACTION?

Use your research question/ title to write your aim. You can start the aim with 'To investigate ...'

AIM

To investigate how temperature can change the rate of chemical reactions

It is a good idea to include the variables in an investigation report. They can help you to write your hypothesis.

- Independent variable: water temperature
- Dependent variable: time taken to complete the reaction
- Controlled variables: amount of water, amount, mass and size of Alka-Seltzer tablets, beaker size

If your investigation is an experiment, then you should include a hypothesis. It should refer to your independent variable and your dependent variable. Remember to write in the third person.

HYPOTHESIS

If the temperature of the water is increased, then the time taken to complete the chemical reaction (dissolve an Alka-Seltzer tablet) will decrease.

List all materials and equipment with amounts and sizes as simple bullet points.

MATERIALS

- 3 alka seltzer tablets
- distilled water at three different temperatures: cold, room temperature and warm (40 °C)
- 3 x 250 mL beakers
- stopwatch
- thermometer

The method provides clear, step-by-step instructions.

Remember to number the steps of your method, and that traditionally methods are written in the past tense and third person. Methods should be written like a cooking recipe – very simple, clear and detailed. A good idea is to imagine that a younger student has to follow your method.

METHOD

1 200 mL each of cold, room temperature and warm water was measured into three separate, labelled beakers.
2 The water temperature of each beaker was measured with the thermometer and recorded in the results table.
3 An Alka-Seltzer tablet was dropped into each beaker, and the stopwatch was used to time how long it took each tablet to dissolve entirely.
4 Steps 1–3 were repeated twice more for a total of three trials per temperature.
5 The time for each beaker was recorded in the results table.

RESULTS

TABLE 1: TIME TAKEN TO DISSOLVE AN ALKA-SELTZER TABLET

Type of water	Water temperature (°C)	Time taken (s)
Cold	4	174
Room temperature	20	110
Warm	40	48

DISCUSSION

As the results in Table 1 show, the Alka-Seltzer dissolved fastest in the warmest water (40 °C), taking a 48 seconds to dissolve. The cold water (4 °C) took the longest, taking 174 seconds to dissolve the tablet.

Use exact figures from your table and compare them to others, to show you have analysed the data.

The results make sense because the warmer the water is, the faster the water particles are moving, and the more collisions there are with the tablet, speeding up the rate of the chemical reaction.

Describe your results (referring to the table or figures) in the discussion and link them to your understanding of science.

One error that may have occurred is that the thermometer may not have been correctly calibrated (an example of a systematic error). Another example of an error is that the tablets may not have been the same size, as they weren't measured. This could be made more accurate by recording the mass of the tablets before adding them to the water to ensure they are all equal. The validity and reliability could be improved by repeating the experiment another two times.

Identify any potential errors here and suggest improvements to the method to try to control these. Discussion questions can be answered here too.

CONCLUSION

The results of this investigation show that water temperature does affect the rate of a chemical reaction. This investigation supported the hypothesis that the warmer the water, the greater the rate of a chemical reaction. This is due to the faster particles in the warm water and the number of collisions they make with the tablet.

Your conclusion summarises the investigation by responding to the aim.

Mention whether the results support or reject your hypothesis, and briefly summarise the investigation, but don't introduce any new information.

REFERENCES

Scientific American, 2020. *Carbonation Countdown: The Effect of Temperature on Reaction Time.* [online] Available at: https://www.scientificamerican.com/article/bring-science-home-carbonation-time/ [Accessed 30 July 2020].

References show the source of any information you used that was not your own.

This is particularly important when you use secondary data.

Many investigation reports also include a background information section at the beginning, and it is important to identify in your references where you found that information.

Investigation **1.1**

Genetic trait survey

KEY SKILL

USING PRIMARY AND SECONDARY SCIENTIFIC EVIDENCE

When you collect your own data it is referred to as primary data; when you obtain data elsewhere (for example, the internet) it is referred to as secondary data. Primary data can be further supported (or refuted) by considering a larger sample size. You could do this by collecting more primary data or by accessing a much larger pool of data online – secondary data. If you are relying on secondary data, it is important to evaluate it to ensure it is valid. You can do this by considering whether the author is credible, whether the source is reputable and whether the information is current and relates to what is being investigated. ***Hint #1:*** *After doing an investigation like this one, which only has a small sample size (your classmates), try to find Australian secondary data for as many traits as you can.*

AIM

To investigate the different genetic traits that exist in a population

MATERIALS

- graph paper
- genetic trait survey
- at least 10 participants

METHOD

1. Copy the results table into your notebook, adding a title and rows as needed.
2. In the second column of the table, record whether you have the genetic trait listed in the first column.
3. Share your results with the rest of the class and create a class tally.
4. Copy the class tally into your table.

QUESTIONS

1. Graph your class results.
2. Which traits were the most prevalent in your class? Which were the least?
3. What other genetic traits could you have tested for?
4. Suggest why some traits are more common than others.
5. Genetic traits are passed down from parents on chromosomes. Describe the structure of a chromosome.

CONCLUSION

Copy and complete:

'The results show that: (*respond to the aim*).'

RESULTS TABLE I1.1

Trait	My response (yes or no)	Total number in class
Curly hair		
Brown eyes		
Attached earlobe		
Cleft chin		
Hitchhiker's thumb		
Ability to roll tongue		
Left-handedness		
Widow's peak hairline		

Investigation **1.2A**

Extracting DNA from strawberries

KEY SKILL

IDENTIFYING AND MANAGING RELEVANT RISKS

Brainstorm with a partner to identify three possible hazards or risks that may be involved in this investigation. Suggest one way that each hazard or risk could be reduced.

AIM

To investigate the process of extracting DNA from strawberries

MATERIALS

- strawberries
- liquid dishwashing detergent
- very cold ethanol (kept in ice or in the freezer)
- 10 mL measuring cyclinder
- plastic spoon
- salt
- water
- ziplock bag
- wooden skewer
- 10 cm squares of gauze
- plastic pipette
- funnel
- large test tube
- ruler
- test-tube rack
- 50–100 mL measuring cylinder
- electronic balance

METHOD

1. Copy the results table into your notebook, adding a title and rows as needed.
2. Make the DNA extraction buffer (this can also be pre-prepared by a lab assistant). Add 5 mL of detergent, 0.75 g of salt (weighed on electronic scales) and 45 mL of water to the larger measuring cylinder. Stir to mix. This will make enough for five 10 mL samples of extraction buffer.
3. Remove the green part of a strawberry and wash the strawberry carefully with water.
4. Place the strawberry in a ziplock bag with 10 mL of the extraction buffer and remove as much air as possible before sealing the bag tightly.
5. On the bench, crush and massage the strawberry inside the bag for a minute.
6. Set up the test tube in a rack on the bench.
7. Place the funnel in the test tube and then line the funnel with the gauze.
8. Pour the mixture from the ziplock bag into the funnel.
9. When the mixture has filtered (this will take some time), discard the gauze and strawberry pulp, keeping only what was filtered into the test tube.
10. Measure the amount of liquid in the test tube very roughly with a ruler.
11. Use the plastic pipette to add the same amount of ice-cold ethanol as the amount of liquid already in the test tube.
12. Observe what happens in the test tube. You should be able to see changes straight away.
13. Use the wooden skewer to gentle remove some DNA from the top of the test tube.

RESULTS

TABLE I1.2A

Observations	Photo or diagram

QUESTIONS

1. What did the strawberry DNA look like?
2. Why do you think it was important to crush the strawberries before extracting the DNA?
3. What do you think the DNA extraction buffer did?

CONCLUSION

Copy and complete:
'The results show that: (*respond to the aim*).'

Investigation **1.2B**

DNA lolly model

KEY SKILL
USING MODELLING AND SIMULATIONS

Scientists can use modelling and simulations to simplify and explain ideas. Models make things easier to understand and visualise, which helps us make predictions. However, there are some limitations to the use of models – how many can you think of?

AIM

To investigate the structure of DNA by making a model out of lollies

MATERIALS

- toothpicks
- red liquorice
- four of the same type of lollies in four colours; e.g. mini-marshmallows, jelly beans, gummy bears
- clean surface or cutting board

METHOD

1. Assign lolly colours to each of the four DNA bases – adenine, cytosine, guanine and thymine.
2. Use two pieces of red liquorice as the DNA backbone.
3. Pair your lolly bases: guanine with cytosine and adenine with thymine.
4. Thread the matching base pairs onto toothpicks.
5. Insert the ends of the toothpicks into the red liquorice DNA backbone.
6. Twist the red liquorice DNA backbone as you go, creating the DNA twisted double helix model.
7. Draw a diagram or take a photo of your finished DNA model.

QUESTIONS

1. Suggest why the use of models is important in science.
2. How could you make your model more accurate? (Hint: Think about what the backbone is made of.)
3. How could you alter this model to show DNA replication?
4. Outline the structure of DNA.
5. Which is bigger – DNA or a gene? Justify your answer.
6. Outline the steps this piece of DNA would go through in order to replicate.
7. What kind of mutations could occur to the DNA in your model?

CONCLUSION

Copy and complete:
'The results show that: (*respond to the aim*).'

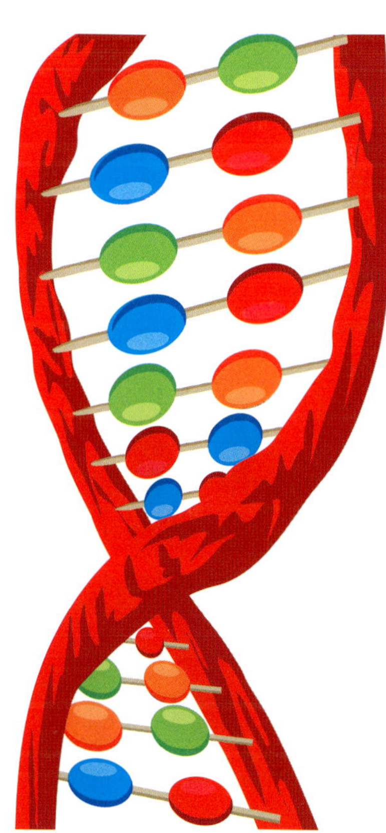

Investigation **2.2**

Natural selection in practice

KEY SKILL

REPRESENTING DATA TO IDENTIFY PATTERNS AND TRENDS

When you write a formal investigation report, there is always a results section that includes your data, often as a table, chart or image. Choosing how to represent your data so that it can be clearly communicated to someone reading your investigation report is an important skill. In this investigation, after you have collected and recorded your data in the results table, turn your data into a chart or a graph. Identify any patterns or trends in your data and include this in your discussion.

Hint #1: *There are many ways to visualise your data, such as bar charts, line graphs and pie charts. Make sure you choose the best one for your data set.*

AIM

To investigate the effectiveness of different 'beaks' in collecting different types of food

MATERIALS

- uncooked rice
- popcorn kernels
- string (cut into 3 cm pieces)
- marbles
- four small plastic tubs
- forceps
- crucible tongs
- kitchen tongs
- blunt pliers
- 4 paper cups
- stopwatch
- graph paper

METHOD

1 Copy the results table into your notebook, adding a title.

2 Place the marbles, rice, pieces of string and popcorn into each of the plastic tubs; that is, marbles in one tub, rice in another tub, etc.

3 Designate a different 'beak' (forceps, crucible tongs, kitchen tongs or blunt pliers) to each person in the group.

4 Choose a person to use the stopwatch to time 2 minutes.

5 Each student uses their tool to pick up as many objects in one of the tubs as possible in the 2 minutes. For example, the student with the forceps attempts to pick up items from the marbles container, the student with the crucible tongs attempts to pick up items from the rice container, etc. Participants place the objects they collect with their tool into their paper cup.

6 After 2 minutes, stop and count how many of each object is in each person's paper cup.

7 Record the totals in your results table.

8 Share your data so that each person can complete their results table.

9 Each student returns their items to the containers.

10 Repeat steps 4–9 three more times. For each turn, students move to the next container to try to pick up a different item using the same instrument.

QUESTIONS

1 What is natural selection?

2 How does this investigation demonstrate the principles of natural selection?

3 Which tools (or beaks) were best at picking up each type of food? Suggest why.

4 What errors could have occurred during this investigation? Suggest ways to prevent them in the future.

CONCLUSION

Copy and complete: 'The results show that: (*respond to the aim*).'

RESULTS TABLE I2.2

Tool	Beak type	Number of pieces of string	Number of grains of rice	Number of popcorn kernels	Number of marbles
Forceps	Long, thin, pointy				
Crucible tongs	Long, medium width, pointy				
Kitchen tongs	Long, wide				
Blunt pliers	Short, thick, wide				

Investigation **2.3**

Environmental and genetic factors affecting survival

KEY SKILL

IDENTIFYING THE VARIABLES AND FORMULATING A HYPOTHESIS

Before you formulate a hypothesis, identify your independent, dependent and controlled variables. The independent variable is the one thing that you purposefully want to change in an investigation. The dependent variable is what you will be measuring. The controlled variables are all the things you need to keep the same throughout the investigation. To formulate your hypothesis, use the following sentence stem: It can be hypothesised that if (something to do with your independent variable), then (something to do with your dependent variable).

Hint #1: *If you get stuck, use the prompts on page 127 to help you.*

Hint #2: *In your conclusion, state whether your hypothesis was supported (correct) or rejected (incorrect).*

AIM

To investigate the effect of colour on the survival of individuals in different habitats

MATERIALS

- 20 red toothpicks
- 20 yellow toothpicks
- 20 green toothpicks
- trundle wheel or tape measure

METHOD

1 Copy the results table into your notebook, adding a title.

2 You will need three different surfaces on which to conduct the investigation: carpet, concrete and grass.

3 Use the trundle wheel or tape measure to measure out a 2 m × 2 m square on the carpet.

4 Select one person in the group to be the predator. Ask the predator to turn away while the toothpicks (the prey) are scattered evenly over the measured square on the carpet.

5 The predator then has 10 seconds to pick up toothpicks. They should pick up only one toothpick at a time.

6 Count how many toothpicks of each colour the predator has picked up and record your results in the table.

7 Repeat steps 3–6 for the concrete and the grass.

8 Calculate the percentage survival rate by dividing the number of survivors by 20 and then multiply that number by 100.

QUESTIONS

1 Which colours and surfaces had the highest survival rate? Suggest why.

2 Why did the predator find it harder to find prey in certain environments?

3 How does this investigation demonstrate the principles of natural selection?

4 What did your results tell you about the role of environmental and genetic factors on survival?

5 What errors could have occurred during this investigation? Suggest ways to ensure they do not occur in the future.

CONCLUSION

Copy and complete:

'The results show that: (*respond to the aim*).'

RESULTS TABLE I2.3

Surface	Toothpick colour	Number of prey caught	Number of survivors	Survival rate (%)
Carpet	Red			
	Yellow			
	Green			
Concrete	Red			
	Yellow			
	Green			
Grass	Red			
	Yellow			
	Green			

Investigation **2.4**

Comparative anatomy

KEY SKILL

EXPLAINING RESULTS USING SCIENTIFIC KNOWLEDGE

When you write a formal investigation report, there is always a discussion section that includes your analysis and explanation of the data you collected. This is where you get to explain your results by linking them to what you already knew about the science of what you are studying.
Hint #1: *You can use the following sentence stem to write about your results: 'My data shows ... and this makes sense because ....'*

AIM

To investigate the comparative anatomy of humans, cats, whales and bats

MATERIALS

- images of human, cat, whale and bat appendages (or full skeletons/models if possible)

METHOD

1. Consider the diagram of the homologous structures of a human, a cat, a whale and a bat.
2. Copy the diagram into your notebook or obtain a photocopy from your teacher.
3. Colour each ulna the same colour on each of the four appendages.
4. Continue to colour the same bones on each appendage in matching colours. Label the bones.
5. Create a Venn diagram in your results section that shows the similarities and differences between each species appendage, based on your observations.

RESULTS

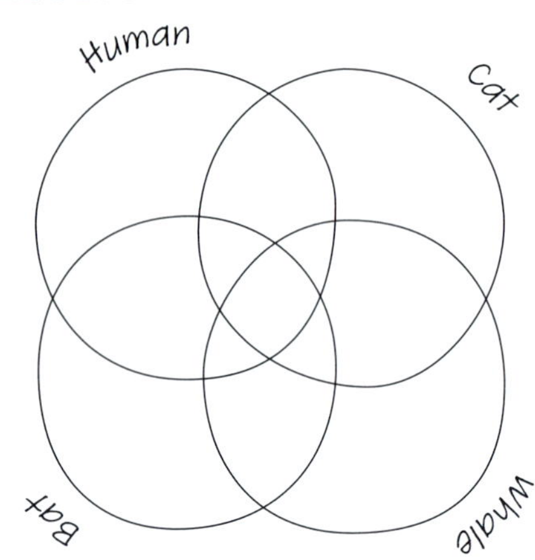

QUESTIONS

1. Identify the bones that are common across all four species.
2. Explain how comparative anatomy can be used as evidence of evolution.
3. Do your results suggest these species had a common ancestor? Why or why not?
4. What does the structure of the bones tell you about the organism and what it uses the appendage for?

CONCLUSION

Copy and complete:
'The results show that: (*respond to the aim*).'

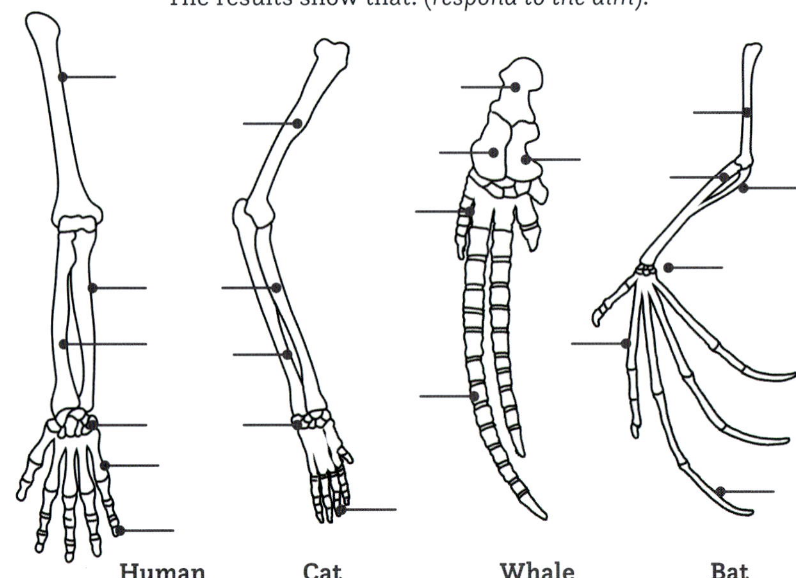

Investigation **2.5**

Dating fossils

KEY SKILL
USING MODELLING AND SIMULATIONS

Scientists can use modelling and simulations to simplify and explain ideas. Models make things easier to understand and visualise, which helps us make predictions. However, there are some limitations to the use of models – how many can you think of?

AIM

To investigate a model for absolute dating

MATERIALS

- 100 five- or ten-cent coins in a container

METHOD

1 Each coin represents an atom in the radioactive element carbon-14. You will be investigating how long the half-life of carbon-14 is. Scientists can calculate the age of fossils by identifying the half-lives of atoms surrounding the fossil.
2 Copy the results table into your notebook, adding a title.
3 Shake the container and carefully empty the coins onto the table. Spread them out so that none overlap.
4 Remove all the coins showing tails. These represent atoms that are no longer radioactive and have decayed.
5 Record the number of coins showing heads in your results table. These represent atoms that are still radioactive.
6 Put the coins showing heads back in the container. Shake the container, and carefully empty the coins onto the table. Repeat this process until all the coins are gone or you have completed 12 trials.
7 Draw a graph of your results.

RESULTS

TABLE I2.5

Shake number	Number of coins showing heads (radioactive atoms)
1	
2	
3	
4	
5	
6	
7	
8	
9	
10	
11	
12	

QUESTIONS

1 How does this model relate to the technique used to identify the age of fossils?
2 What is the difference between relative and absolute dating?
3 How many shakes did it take for half of your coins to decay? This represents the half-life of your carbon-14.
4 The half-life of carbon-14 is actually 5730 years. If each coin shake is 5730 years, how many years would it take for carbon-14 to completely disappear?
5 What did your results tell you about absolute dating and the decay of radioactive isotopes?
6 What errors could have occurred during this investigation? Suggest ways to ensure they do not occur in the future.

CONCLUSION

Copy and complete:
'The results show that: (*respond to the aim*).'

Investigation **3.4A**

Physical properties of elements

KEY SKILL

EXPLAINING RESULTS USING SCIENTIFIC KNOWLEDGE

When you write a formal investigation report, there is always a discussion section that includes your analysis and explanation of the data you collected. This is where you get to explain your results by linking them to what you already knew about the science of what you are studying.

Hint #1: *You can use the following sentence stem to write about your results: 'My data shows ... and this makes sense because ...'.*

AIM

To investigate and compare the physical properties of metals and non-metals

MATERIALS

- samples of various metals and non-metals, such as lead, zinc, aluminium, iron, magnesium, tin, sulfur and carbon
- power-pack, switch and electrical leads
- light bulb
- sand paper
- heatproof mat

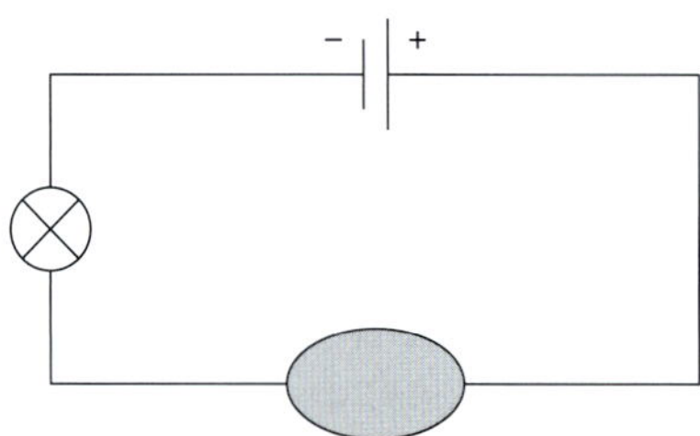

Tested material

METHOD

1. Copy the results table into your notebook, adding a title and rows as required.
2. Clean and polish each sample with sand paper and record the appearance after it's been cleaned. Record your observations.
3. Test for malleability by trying to bend the sample. Record your observations.
4. Using a power-pack, light bulb, switch and cables, construct a circuit and test each sample for electrical conductivity on the heatproof mat. Check if the bulb lights up, and record your observations.
5. Using a periodic table and your observations, determine whether the element you have tested is a metal or a non-metal.

QUESTIONS

1. List the physical properties of metals that you observed.
2. List the physical properties of non-metals that you observed.
3. Predict, using the elements tested and their properties, which other elements would have similar properties.

CONCLUSION

Copy and complete:

'The results show that: (*respond to the aim*).'

RESULTS TABLE I3.4A

Element	Appearance when polished	Bends or crumbles	Electrical conductivity	Metal or non-metal

Investigation **3.4B**

Chemical properties of elements

KEY SKILL
WRITING A RESEARCH QUESTION

Turn the aim of this investigation into a question that asks what you are trying to discover. This is called a research question.
Hint #1: *Make sure that your research question has a question mark at the end.*
Hint #2: *Your research question can also be used as a title for an experiment report.*

AIM

To investigate and compare the chemical properties of metals and non-metals

MATERIALS

- 1 mol/L hydrochloric acid
- water
- 1 cm magnesium ribbon cut into three pieces
- iron filings
- copper turnings
- sulfur powder
- 4 test tubes
- 4 gas jars filled with oxygen
- 4 deflagrating spoons
- Bunsen burner
- heatproof mat
- matches
- safety glasses
- heatproof gloves

Deflagrating spoon
Gas jar containing oxygen
Element sample
Water

WEAR SAFETY GLASSES AND HEATPROOF GLOVES DURING HEATING. HEAT THE ELEMENTS WITH THE DEFLAGRATING SPOON. ENSURE YOU HEAT THE SULFUR IN A FUME CUPBOARD. DO NOT LOOK DIRECTLY AT THE FLAME.

METHOD

1. Copy the results table into your notebook, adding a title and rows as required.
2. Put a small amount of copper turnings in a test tube. Add about 3 mL of hydrochloric acid. Record your observations.
3. Repeat step 2 with magnesium, iron filings and sulfur powder.
4. Place a small amount of copper in a deflagrating spoon and heat it. Then place it in a gas jar containing oxygen gas and a small amount of water. Record your observations. (Do not look directly into the flame.)
5. Repeat step 4 with the magnesium and iron filings. Record your observations.
6. Repeat step 4 with a small amount of sulfur. This must be performed in the fume cupboard.

RESULTS

TABLE 13.4B

Element	Reaction with acid	Reaction with oxygen

QUESTIONS

1. Describe the differences between the effects of acids on metals and non-metals.
2. Describe the effects of reacting metals and non-metals with oxygen.
3. Give reasons why the elements tested are in particular locations in the periodic table.

CONCLUSION

Copy and complete:
'The results show that: (*respond to the aim*).'

Investigation **4.5A**

The effect of concentration on the rate of a reaction

KEY SKILL

IDENTIFYING AND MANAGING RELEVANT RISKS

Brainstorm with a partner to identify three possible hazards or risks that may be involved in this investigation. Suggest one way that each hazard or risk could be reduced.

AIM

To investigate how concentration affects the rate of chemical reactions

MATERIALS

- 0.15 mol/L sodium thiosulfate
- 2 mol/L hydrochloric acid
- 5 × 100 mL beakers
- 2 × 50 mL measuring cylinders
- 5 × 10 mL measuring cylinders
- black whiteboard marker pen
- stirring rod
- distilled water
- stopwatch

METHOD

1. Copy the results table into your notebook, adding a title.
2. Label the five beakers 1–5.
3. Mark the bottom of each beaker with an X.
4. Use the two 50 mL measuring cylinders to add sodium thiosulfate and distilled water in the volumes indicated in the results table.
5. Add 5 mL of hydrochloric acid solution to each 10 mL measuring cylinder and place one next to each beaker.
6. Carefully add the hydrochloric acid to the sodium thiosulfate solution in beaker 1.

   Stir the solution once with a stirring rod and immediately start timing.
7. Stop timing when the black X is no longer visible.
8. Record the reaction time in seconds in the results table.
9. Wash the stirring rod between each use. Repeat steps 6–9 with the remaining beakers.

   Calculate $\frac{1}{\text{reaction time}}$ for each trial.

   Plot concentration vs time and concentration vs $\frac{1}{\text{time}}$ on separate graphs.

DISCUSSION

1 List the controlled variables, independent variable and dependent variable.
2 Should the total volume of each beaker be the same or different? Explain your answer?

CONCLUSION

Copy and complete:
'The results show that: (*respond to the aim*).'

SODIUM THIOSULFATE IS HARMFUL IF INGESTED. DO NOT INGEST IT. WASH YOUR HANDS AFTER USING IT. HYDROCHLORIC ACID IS CORROSIVE TO EYES AND SKIN AND IS AN IRRITANT. IT MAY CAUSE BURNS. IF SPILT ON SKIN FLUSH THE AREA WITH COLD TAP WATER. WEAR SAFETY GLASSES. SULFUR DIOXIDE IRRITATES THE RESPIRATORY SYSTEM. MAKE SURE THE ROOM IS VENTILATED.

RESULTS TABLE I4.5A

Beaker	Sodium thiosulfate (mL)	Water (mL)	Sodium thiosulfate (mol/L)	Time (s)	$\frac{1}{time}$ (s^{-1})
1	50	0	0.15		
2	40	10	0.12		
3	30	20	0.090		
4	20	30	0.060		
5	10	40	0.030		

Investigation **4.5B**

The effect of temperature on the rate of a reaction

KEY SKILL

REPRESENTING DATA TO IDENTIFY PATTERNS AND TRENDS

When you write a formal investigation report, there is always a results section that includes your data, often as a table, chart or image. Choosing how to represent your data so that it can be clearly communicated to someone reading your investigation report is an important skill. In this investigation, after you have collected and recorded your data in the results table, turn your data into a chart or a graph. Identify any patterns or trends in your data and include this in your discussion.

Hint #1: *There are many ways to visualise your data, such as bar charts, line graphs and pie charts. Make sure you choose the best one for your data set.*

AIM

To investigate the effect of temperature on the rate of a reaction

MATERIALS

- 3 Alka-Seltzer tablets
- distilled water at three different temperatures: cold, room temperature and warm (40°C)
- 3 × 250 mL beakers
- stopwatch
- thermometer

METHOD

1. Copy the results table into your notebook, adding a title.
2. Measure 200 mL each of warm water, room temperature water and cold water into three separate labelled beakers.
3. Measure the water temperature in each beaker and record it in the results table.
4. Drop an Alka-Seltzer tablet into each beaker. Use the stopwatch to immediately start timing how long it takes to dissolve entirely.
5. Stop the timer when the tablet has dissolved completely.
6. Record the reaction times (in seconds) in the results table.

RESULTS

TABLE I4.5B

Type of water	Temperature (°C)	Time to dissolve (s)
Cold		
Room temperature		
Warm		

DISCUSSION

1 Graph the reaction time (seconds) on the *y*-axis against the water temperature (°C) on the *x*-axis.
2 List the controlled variables, independent variable and dependent variable.
3 How does reaction time change with temperature? Do the results support your hypothesis? Why or why not?
4 If some tablets used are whole and some are broken, would the temperature experiment still be valid? Why or why not?
5 If some students stirred the water while their tablets were dissolving, would the temperature experiment still be valid? Why or why not?
6 If everyone in the class used a different water temperature, could you average the results? Why or why not?

CONCLUSION

Copy and complete:
'The results show that: (*respond to the aim*).'

WEAR SAFETY GLASSES WHILE DOING THIS EXPERIMENT TO PREVENT ANY MATERIAL FROM ENTERING YOUR EYES.

Investigation **4.6A**

The effect of a catalyst on the rate of a reaction

KEY SKILL
DRAWING CONCLUSIONS CONSISTENT WITH EVIDENCE

When you write a formal investigation report, there is always a conclusion section that summarises the investigation by responding to (or answering) the aim. To do this you need to draw a conclusion that is consistent with the data or evidence you collected.

Hint #1: *You can use the following sentence stem in your conclusion: 'The results of this investigation show ...'.*

Hint #2: *Make sure your conclusion answers or responds to your aim.*

AIM

To investigate the effect of a catalyst on the decomposition of hydrogen peroxide

Hydrogen peroxide solution decomposes slowly at room temperature to water and oxygen:

hydrogen peroxide	→	oxygen	+	water
$2H_2O_2(aq)$	→	$O_2(g)$	+	$2H_2O(l)$

The volume of oxygen given off can be measured by collecting it in an inverted measuring cylinder.

MATERIALS

- 3% hydrogen peroxide
- 0.5 g manganese dioxide
- conical flask with a delivery tube attached with a rubber stopper
- 50 mL and 100 mL measuring cylinders
- retort stand, boss head and clamp
- trough of water

METHOD

1. Copy the results table into your notebook, adding a title.
2. Using a 50 mL measuring cylinder, add 50 mL of hydrogen peroxide solution to a conical flask.
3. Fill the 100 mL measuring cylinder with water, seal the top with your fingers, and invert it into the water trough, as shown in the diagram.
4. Put the rubber stopper into the conical flask and make sure the delivery tube connects to the inverted measuring cylinder.
5. Record the volume of gas given off every 10 seconds for 60 seconds.
6. Measure 0.5 g of the catalyst manganese dioxide.
7. Add the catalyst to the conical flask, put the stopper loosely back into the flask and start the stopwatch.
8. Record the volume of gas given off every 10 seconds. Continue timing until no more oxygen appears to be given off. Record all your observations.

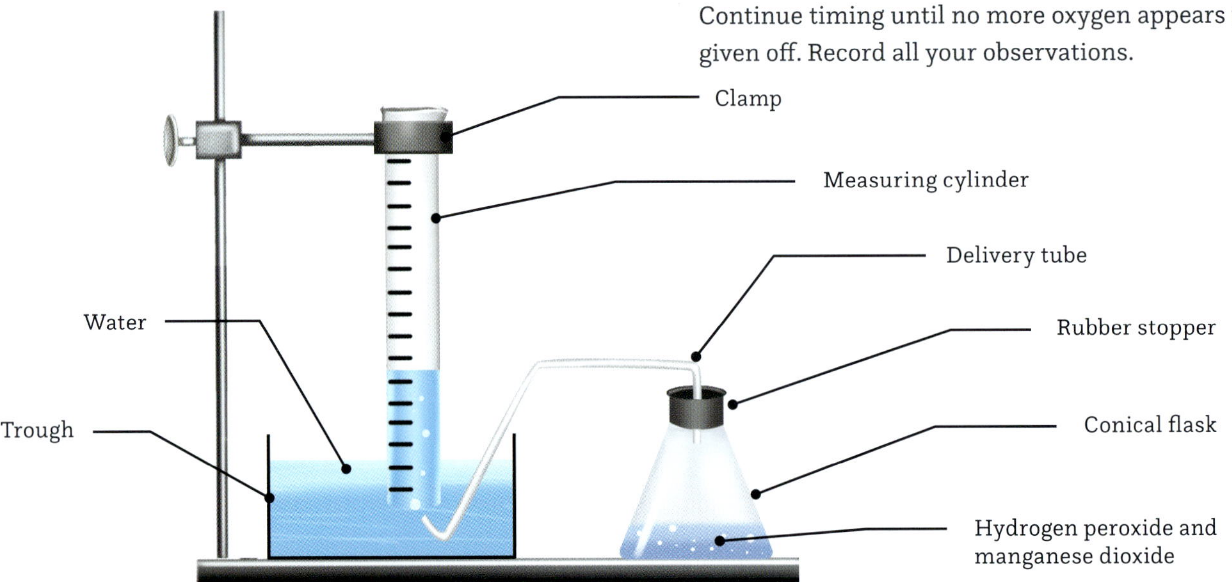

RESULTS

TABLE I4.6A

Time (s)	Volume of $O_2(g)$ (mL) with catalyst	Volume of $O_2(g)$ (mL) without catalyst
10		
20		
30		
40		
50		
60		

DISCUSSION

1. Draw a graph of your results with time on the *x*-axis and volume of oxygen on the *y*-axis.
2. Comment on the trends you see on the graphs.
3. Identify the catalyst in this reaction.
4. What was the role of the catalyst in this experiment?
5. Did the temperature in the conical flask change?

CONCLUSION

Copy and complete:

'The results show that: (*respond to the aim*).'

HYDROGEN PEROXIDE CAN CAUSE SKIN IRRITATION. WEAR GLOVES AND WASH YOUR HANDS IMMEDIATELY IF HYDROGEN PEROXIDE GETS ONTO THEM.

MANGANESE DIOXIDE CAN IRRITATE IF INHALED, SO AVOID SPILLING IT.

DISPOSE OF WASTES APPROPRIATELY.

Investigation **4.6B**

The effect of surface area on the rate of a reaction

KEY SKILL

IDENTIFYING THE VARIABLES AND FORMULATING A HYPOTHESIS

Before you formulate a hypothesis, identify your independent, dependent and controlled variables. The independent variable is the one thing that you purposefully want to change in an investigation. The dependent variable is what you will be measuring. The controlled variables are all the things you need to keep the same throughout the investigation. To formulate your hypothesis, use the following sentence stem: It can be hypothesised that if (something to do with your independent variable), then (something to do with your dependent variable).

Hint #1: *If you get stuck, use the prompts on page 127 to help you.*

Hint #2: *In your conclusion, state whether your hypothesis was supported (correct) or rejected (incorrect).*

AIM

To investigate the effect of increasing the surface area on the rate of a reaction

MATERIALS

- 4 Alka-Seltzer tablets
- water
- 4 × 100 mL beakers
- mortar and pestle
- 50 mL measuring cylinder

METHOD

1. Copy the results table into your notebook, adding a title.
2. Add 50 mL of water to each of the four beakers.
3. Of the four Alka-Seltzer tablets:
   - leave the first one whole
   - cut the second one in half
   - cut the third one into quarters
   - crush the fourth one with the mortar and pestle.
4. Add the whole tablet to the first beaker of water and time how long it takes to dissolve.
5. Add the pieces of the second tablet to the second beaker of water and time how long they take to dissolve.
6. Add the pieces of the third tablet to the third beaker of water and time how long they take to dissolve.
7. Add the crushed tablet to the fourth beaker of water and time how long it takes to dissolve.
8. Make your observations and record your results.

RESULTS **TABLE I4.6B**

Size of Alka-Seltzer tablet	Time taken for the tablet to dissolve (s)
Whole tablet	
Halved tablet	
Quartered tablet	
Crushed tablet	

QUESTIONS

1. List the controlled variables, independent variable and dependent variable.
2. Do you think your experiment was reliable? Why or why not?
3. Draw a column graph to show your results.
4. Explain what you think the rate of reaction depended on.
5. Explain how you changed the surface area of the tablets.

CONCLUSION

Copy and complete:
'The results show that: (*respond to the aim*).'

WEAR SAFETY GLASSES WHILE DOING THIS EXPERIMENT TO PREVENT ANY PARTICLES FROM ENTERING YOUR EYES.

Investigation **5.1**

Observing interactions between the spheres

KEY SKILL

WRITING A RESEARCH QUESTION

Turn the aim of this investigation into a question that asks what you are trying to discover. This is called a research question.

Hint #1: *Make sure that your research question has a question mark at the end.*

Hint #2: *Your research question can also be used as a title for an experiment report.*

AIM

To investigate different interactions between the atmosphere, lithosphere, hydrosphere and biosphere

MATERIALS

- camera

METHOD

1. Copy the results table into your notebook, adding rows as required.
2. Find a space outside. This can be a natural or manufactured environment.
3. Sit and quietly observe your surroundings.
4. Record any interactions you can clearly observe between two or more of the four Earth spheres.

QUESTIONS

1. Describe the location where you made your observations.
2. What sphere(s) was easiest to observe? Why do you think this was the case?
3. What sphere(s) was hardest to observe? Why do you think this was the case?
4. Was there an observation you made that involved all of the spheres? If not, what was the maximum number of spheres that you observed interacting? Explain what the interaction was and how all of the spheres were involved.

CONCLUSION

Copy and complete:

'The results show that: (*respond to the aim*).'

RESULTS TABLE I5.1

Observation	Spheres that are involved				Explanation
	Atmosphere	Lithosphere	Hydrosphere	Biosphere	

Investigation **5.2A**

Releasing dinosaur breath

KEY SKILL

EXPLAINING RESULTS USING SCIENTIFIC KNOWLEDGE

When you write a formal investigation report, there is always a discussion section that includes your analysis and explanation of the data you collected. This is where you get to explain your results by linking them to what you already knew about the science of what you are studying.

Hint #1: *You can use the following sentence stem to write about your results: 'My data shows ... and this makes sense because ...'.*

AIM

To investigate the reaction between chalk and dilute hydrochloric acid

MATERIALS

- crushed chalk
- 0.1 mol/L hydrochloric acid
- 25 mL limewater (diluted calcium hydroxide)
- 100 mL conical flask
- 25 mL measuring cylinder
- boiling tube
- test tube rack
- balloon

METHOD

1. Half-fill the boiling tube with limewater and place it in the rack.
2. Carefully add the crushed chalk to the conical flask.
3. Use the measuring cylinder to add 25 mL of 0.1 mol/L hydrochloric acid to the chalk.
4. Quickly and carefully place the opening of the balloon over the neck of the flask.
5. Once the reaction has slowed and the balloon is no longer inflating, pinch the balloon tightly around the neck as you remove it from the flask.
6. Place the neck of the balloon over the boiling tube and squeeze the balloon to force the gas in.

QUESTIONS

1. Chalk is the remains of tiny marine organisms with shells made of calcium carbonate ($CaCO_3$). Use your knowledge of the carbon cycle to identify where the carbon in the chalk came from. Where was it before then?
2. Write the word equation for the chemical reaction that occurred between the chalk and the acid.
3. Why could you say that 'dinosaur breath' was released from the chalk? Draw a diagram of the carbon cycle to help illustrate your response.

CONCLUSION

Copy and complete:

'The results show that: (*respond to the aim*).'

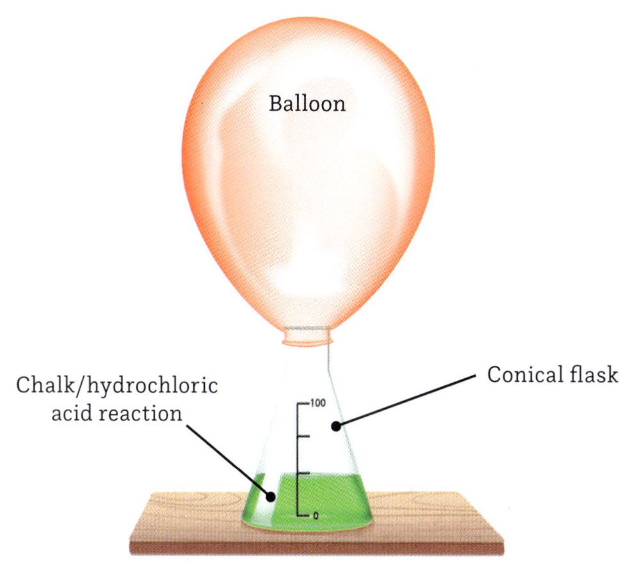

Investigation **5.2B**

The effect of temperature on soil respiration

KEY SKILL

IDENTIFYING THE INDEPENDENT, DEPENDENT AND CONTROLLED VARIABLES

The independent variable is the one thing that you purposefully want to change in an investigation.
The dependent variable is what you will be measuring.
The controlled variables are all the things you need to keep the same throughout the investigation.
Hint #1: *Brainstorm with a partner three things that will be or were kept the same in your investigation. These will be the controlled variables.*

AIM

To determine how temperature affects the respiration rate (and therefore the rate that carbon is released) of soil microbes

MATERIALS

- well-dried topsoil
- 1% sugar water solution
- 150 mL limewater (diluted calcium hydroxide)
- 3 × 600 mL plastic bottles
- 3 × 100 mL conical flasks
- 3 one-hole stoppers with glass tubes, to fit bottles
- 3 one-hole stoppers with glass tubes, to fit conical flasks
- 3 pieces of rubber tubing
- 25 mL measuring cylinder
- 50 mL measuring cylinder
- thermometer
- tape
- scissors
- safety blade
- small trowel or scoop
- marker pen
- ruler
- heat lamp

METHOD

1. Copy the results table into your notebook, adding a title.
2. Place a stopper in the mouth of a bottle. Use a ruler to measure and the pen to mark 1 cm from the base of the stopper. Remove the stopper.
3. Measure 4 cm from the top of the bottle and mark. Use the safety blade to cut about 80% of the way around the bottle, at the 4 cm mark. Carefully bend back the top.
4. Fill the bottle approximately one-third full with soil then use the measuring cylinder to add 20 mL of 1% sugar solution. Gently tap the bottle on the bench to compact and settle the soil.
5. Add more soil until the bottle is two-thirds full and use the measuring cylinder to add 20 mL of 1% sugar solution. Gently tap the bottle on the bench.
6. Fill the bottle with soil up to the cut. Use the measuring cylinder to add 20 mL of 1% sugar solution. Gently tap the bottle on the bench.
7. Flip the top back into place and seal the cut with tape.
8. Carefully add more soil until it reaches the mark 1 cm from the base of the stopper. Use the measuring cylinder to add 10 mL of 1% sugar solution. Gently tap the bottle on the bench.
9. Insert the stopper with glass tubing.
10. Repeat steps 2–9 for the other bottles.
11. Use a measuring cylinder to add 50 mL of limewater to each of the conical flasks.
12. Insert the stoppers with glass tubing into the conical flasks, ensuring the tubing is in the limewater.
13. Connect one soil bottle to each of the conical flasks with the rubber tubing.
14. Place one apparatus under a heat lamp.
15. Place the second apparatus in a fridge.
16. Place the third apparatus in the classroom.
17. Record the air temperatures for each location.
18. Record your observations of the colour of the limewater over the next 3–5 days.

QUESTIONS

1 Compare the colour in the limewater across the three flasks. What does a change in colour in the limewater indicate?

2 Microorganisms in the soil consume organic material, taking in carbon. They then release this carbon as carbon dioxide when they undergo cellular respiration. Compare the rates of cellular respiration in each flask, identifying which flask had the least and the most.

3 Describe the relationship that you observed between temperature and soil microbe respiration rate. Was this expected? Why or why not?

4 Where does the carbon dioxide released by soil microbes usually go?

5 What respiration rate would you expect to observe in soils in a region that experiences a:

a very cold climate?

b tropical climate?

6 Is there any evidence from your investigation that a warming climate could change soil microbe respiration rates?

DISCUSSION

1 Identify the dependent, independent and controlled variables in this investigation.

2 Were there any variables that were not controlled? How could have they affected the results? Propose changes to the procedure to ensure that these variables are controlled.

CONCLUSION

Copy and complete:

'The results show that: (*respond to the aim*).'

RESULTS **TABLE 15.2B**

Apparatus location	Temp. (°C)	Observations
Heat lamp		
Fridge		
Classroom		

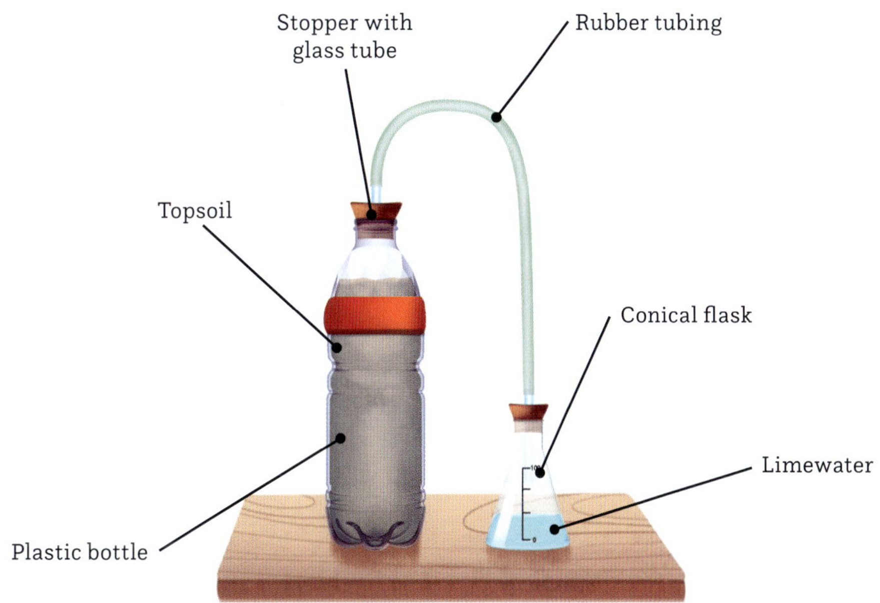

Investigation **5.3**

Investigating sunscreens

KEY SKILL

EVALUATING RESULTS FOR RELIABILITY AND VALIDITY

In order for our investigations to be considered scientific, we need to check that our results were reliable and valid. It sounds like a difficult thing to check but it's actually simple. If your results are reliable, it means that if you repeated your test or investigation you would get the same results. If your results are valid, it means that you were able to measure what was intended to be measured.

Hint #1: *If someone makes a human error (for example, dropping something, adding too much or too little of a substance, spilling something or using different equipment each time) then the results are probably not valid or reliable.*

AIM

To determine which material best blocks UV radiation

MATERIALS

- 5 sheets of sun-sensitive paper
- at least four different sun-protection materials (e.g. sunscreen, zinc cream, sunglasses, fabric)
- 5 ziplock bags
- tape
- stopwatch
- camera

METHOD

1. In a darkened room and away from UV light, label the back of a piece of sun-sensitive paper 'Control' and place it inside a ziplock bag.
2. Label the back of a second piece of sun-sensitive paper 'Sunscreen' and place it inside the ziplock bag. Apply the sunscreen to the front of the ziplock bag.
3. Repeat step 3 for the other sun-protection materials being tested, applying the cream or taping the material to the front of the ziplock bag.
4. Carefully cover the bags while you transport them outside.
5. Quickly set them up in direct sunlight and start the stopwatch.
6. Leave the bags for 5 minutes or as per the instructions on the sun-sensitive paper.
7. Bring the bags back inside.
8. Remove the paper from the bags and follow the instructions on the sun-sensitive paper to fix the image.
9. Photograph the results or record your observations.

QUESTIONS

1 Compare the papers that were 'protected' to the control. Were there significant differences?

2 a Which material was the most effective at blocking the UV radiation?

 b Which material was the least effective at blocking the UV radiation?

 c Why do you think this was the case?

3 Why is it important to block as much UV radiation as possible from hitting the skin?

4 Propose what you would expect to see from your results if the Montreal Protocol was not enacted.

5 Design an investigation that compares variations of one type of material; for example, different sunscreen brands, amount of sunscreen applied or different fabrics.

CONCLUSION

Copy and complete:

'The results show that: (*respond to the aim*).'

Investigation **5.4**

The greenhouse effect

KEY SKILL

REPRESENTING DATA TO IDENTIFY PATTERNS AND TRENDS

When you write a formal investigation report, there is always a results section that includes your data, often as a table, chart or image. Choosing how to represent your data so that it can be clearly communicated to someone reading your investigation report is an important skill. In this investigation, after you have collected and recorded your data in the results table, turn your data into a chart or a graph. Identify any patterns or trends in your data and include this in your discussion.

Hint #1: *There are many ways to visualise your data, such as bar charts, line graphs and pie charts. Make sure you choose the best one for your data set.*

AIM

To investigate how a greenhouse affects air temperature

MATERIALS

- 3 glass microscope slides
- sticky tape
- 2 thermometers
- stopwatch
- Blu-Tack/plasticine

METHOD

1. Copy the results table into your notebook, adding a title and rows as needed.
2. Use the sticky tape to stick the long edges of the microscope slides together to form a triangular prism. This is your greenhouse.
3. Use the tape to seal shut one of the open ends.
4. Place the bulb of a thermometer through the other end of the prism before sealing the end with tape. Use Blu-Tack to support the thermometer so that the bulb does not touch the glass slides.
5. Place your greenhouse outside in direct sunlight. Place the second thermometer next to the greenhouse, using Blu-Tack or plasticine to support it. Make sure the thermometer bulb does not touch any surface.
6. Record the initial temperature of both thermometers.
7. Record the temperature of both thermometers every minute for 20 minutes.
8. Quickly cover or remove the greenhouse and thermometers from sunlight.
9. Continue to record the temperature every minute for a further 20 minutes.

RESULTS TABLE I5.4

In sunlight			Out of sunlight		
Time (min)	Temperature (°C)		Time (min)	Temperature (°C)	
	Internal greenhouse	External to greenhouse		Internal greenhouse	External to greenhouse

QUESTIONS

1 Create a graph of your results.
2 Describe the trend shown on your graph.
3 Describe the difference between the temperature in the greenhouse and the outside air temperature when placed in sunlight to warm.
4 a Describe the difference between the temperature in the greenhouse and the outside air temperature out of the sunlight.
  b Explain how this is similar to the greenhouse effect caused by greenhouse gases.
5 Use the data gathered in this investigation to explain why horticulturists would grow plants inside greenhouses as opposed to in the open air.

DISCUSSION

1 Identify the independent, dependent and controlled variables in this investigation.
2 Identify any variables that should have been controlled. How would these have affected the results? Suggest changes to the method that would have ensured these were controlled.

CONCLUSION

Copy and complete:
'The results show that: (*respond to the aim*).'

Investigation **6.1**

Modelling the expanding universe

KEY SKILL

USING MODELLING AND SIMULATIONS

Scientists can use modelling and simulations to simplify and explain ideas. Models make things easier to understand and visualise, which helps us make predictions. However, there are some limitations to the use of models – how many can you think of?

AIM

To determine how a galaxy's distance from a reference point affects its speed

MATERIALS

- a round balloon
- 5 different coloured stick-on dots
- piece of string approximately 50 cm
- ruler
- stopwatch

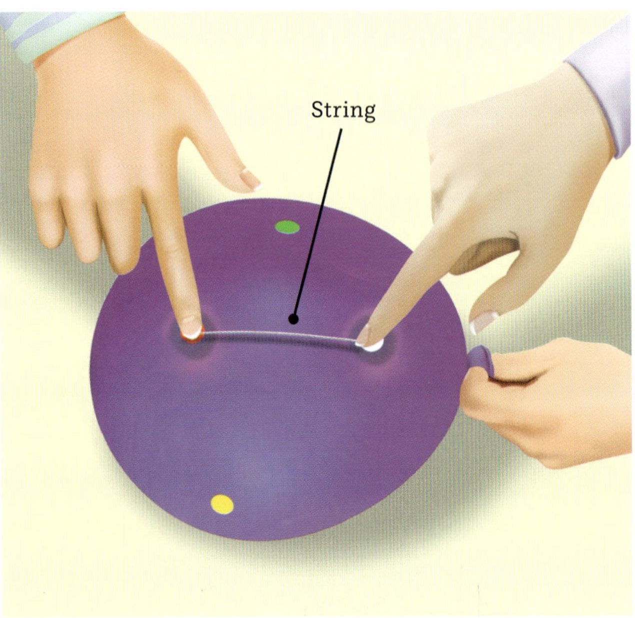

METHOD

1. Copy the results table into your notebook, adding a title.
2. Blow up the balloon to about 150 mm and hold the nozzle closed (do not tie it up).
3. Stick the 5 dots (galaxies) onto the balloon (universe). Try to spread them evenly around the balloon.
4. Select one of the dots to be your home galaxy. Use the string to measure the distance between your home galaxy and the other galaxies, as shown in the diagram. Measure the string with the ruler to determine this distance in centimetres.
5. Record these distances in the results table. (Note: The distance from your home galaxy will be 0 cm.)
6. Fully inflate the balloon and use the stopwatch to time how long it takes. Tie the balloon and record the time taken in the results table.
7. Use the string and ruler to again measure the distance from your home galaxy to the other galaxies. Record these distances in the results table.
8. Calculate the change in distance by subtracting the first measurements from the second measurements.
9. Calculate the speed of each galaxy by dividing the change in distance (cm) by the time it took the balloon to inflate (s). (Note: The speed of your home galaxy will be 0 cm/s.)
10. Plot a line graph with speed (cm/s) on the y-axis and distance (cm) on the x-axis.

RESULTS

TABLE 16.1

Colour of dot	Distance from home galaxy before inflating (cm)	Distance from home galaxy after inflating (cm)	Change in distance (cm)	Speed $\left(\frac{\text{distance}}{\text{time}}\right)$ (cm/s)
Red				
Green				
Blue				
White				
Yellow				
Time to fully inflate the balloon (s)				

QUESTIONS

1. a Are the speeds of all the galaxies the same?
   b If not, what is the relationship between speed and the distance from the home galaxy?
2. Suggest what would happen to the results if you had used a different home galaxy.
3. Explain what the results of this investigation tell you about the way that the universe is expanding.
4. Use the results from this investigation to explain why we observe light from distant galaxies more redshifted than those closer to us.

CONCLUSION

Copy and complete:

'The results show that: (*respond to the aim*).'

Investigation **6.2**

Stargazing

KEY SKILL

REFERENCING SOURCES OF INFORMATION

As your science skills become more advanced, you may wish to do some research prior to completing an investigation. This allows you to understand the investigation better; in particular, the science of what is happening. Whenever you do research it is important to get information from trusted sources and to reference where the information came from. When you reference a source, you include details such as who the author of the information is, when it was published and the title of the website or article.

Hint #1: *Two widely used referencing conventions are Harvard and APA. You can look these up to learn more about them, or there are even websites that will format your references for you.*

AIM

To investigate the night sky

MATERIALS

- star map for the month and year you are viewing, or a stargazing app
- blank A4 paper
- pencil and highlighter
- red torch or torch with a red cellophane cover
- compass
- binoculars or telescope (optional)

METHOD

PART 1

EXPLORING THE NIGHT SKY

1 On a clear night, find a dark area where you will have a good view of the night sky.

2 Use the red torch to help you see your papers.

3 Use the compass to identify north, south, east and west in your location.

4 Orientate your star map.

5 Use your star map to identify:

  a the Southern Cross and Pointers

  b at least two other constellations

  c a planet (if visible at the time you are observing).

6 Use the highlighter to highlight the objects on your sky map that you locate.

7 Compare what you observe with the naked eye to what you can see with a telescope or binoculars.

PART 2

LOCATE SOUTH USING THE SOUTHERN CROSS AND POINTERS

1 Draw an imaginary line with your finger from the top of the cross to the bottom and extending beyond it, as shown on the diagram.

2 Draw an imaginary line joining the two Pointers. Midway along this line, draw another line at right angles to it.

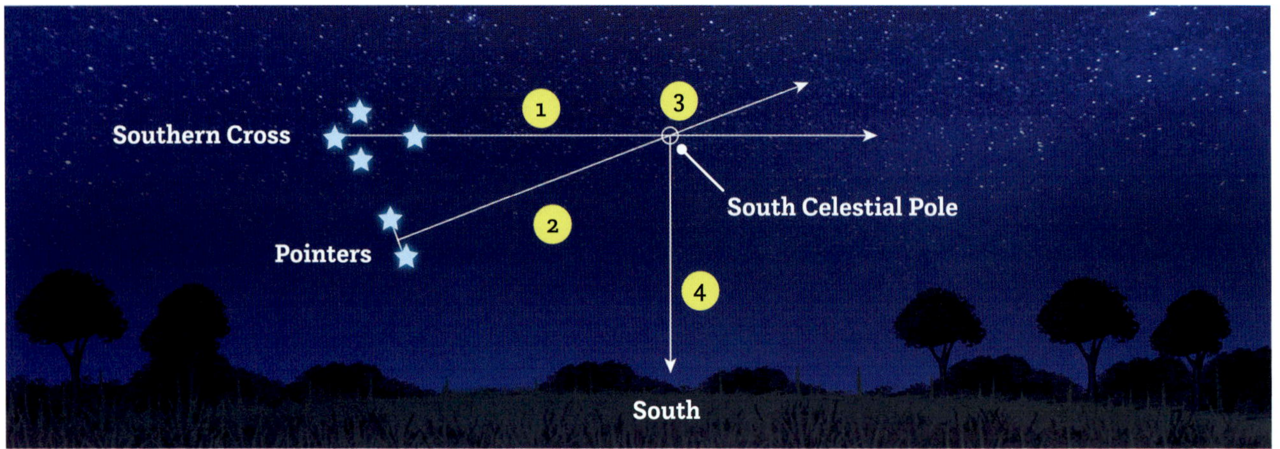

3 Where lines 1 and 2 meet is the South Celestial Pole. This is the point in the sky around which all the stars seen from the Southern Hemisphere rotate.
4 Locate south by dropping a vertical line from the South Celestial Pole to the horizon.

PART 3
OBSERVING THE MOTION OF THE STARS

1 Identify a bright star that will be easy for you to identify again. It should be close to an object such as a roof, tree or chimney.
2 Take note of where you are standing so that you can return to it later.
3 On your piece of paper, draw a silhouette of the object that you are comparing your bright star to, as shown in the diagram.
4 Mark the location of the bright star on your paper. Make a note of the time you make your observation.
5 For the period of your observations, return to the same spot every 30 minutes and note the location of the star.

QUESTIONS

PART 1

1 If you used binoculars or a telescope, describe the differences between the observations you made with them and those you made with the naked eye.

PART 2

2 Compare the technique using the Southern Cross to locate south with using the compass.

PART 3

3 Describe the movement of the star as time passed.
4 Identify the point that the movement is occurring around. Add this to your diagram if appropriate.
5 What causes this movement?

CONCLUSION

Copy and complete:
'The results show that: (*respond to the aim*).'

Investigation **6.4**

Investigating orbits

KEY SKILL

REPRESENTING DATA TO IDENTIFY PATTERNS AND TRENDS

When you write a formal investigation report, there is always a results section that includes your data, often as a table, chart or image. Choosing how to represent your data so that it can be clearly communicated to someone reading your investigation report is an important skill. In this investigation, after you have collected and recorded your data in the results table, turn your data into a chart or a graph. Identify any patterns or trends in your data and include this in your discussion.

Hint #1: *There are many ways to visualise your data, such as bar charts, line graphs and pie charts. Make sure you choose the best one for your data set.*

AIM

To determine the relationship between orbit radius and orbital period

MATERIALS

- glass or plastic tubing 10–20 cm long
- 50 g mass
- mass carrier (50 g)
- string
- ruler
- rubber stopper
- stopwatch
- scissors
- marker pen

METHOD

1. Copy the results table into your notebook, adding a title.
2. Measure and cut 110 cm of string.
3. Tie one end of the string to the rubber stopper.
4. Thread the string through the tubing.
5. Tie the 50 g mass carrier to the other end of the string. Make sure there is 100 cm between the mass and the stopper.
6. Use the ruler and the marker to mark 10 cm intervals between the mass and the stopper.
7. With the stopper at a 50 cm distance from the tubing, hold the tube above your head and swing the stopper around until it is spinning steadily. The 50 g mass should hang in the same position. If it doesn't, you will need to increase or decrease the speed of spinning.
8. Use the stopwatch to time how long it takes for the stopper to complete 10 orbits of the tubing.
9. Record your results in the table and calculate the time taken for one orbit.
10. Repeat steps 7–9 for the other distances.
11. Draw a line graph of your results.

RESULTS **TABLE 16.4**

Orbit radius (cm)	Time to complete 10 orbits (s)	Time to complete 1 orbit (s)
50		
40		
30		
20		
10		

Planet orbit distances and time to orbit

Planet	Distance from Sun (AU)	Time to orbit (Earth years)
Mercury	0.4	0.2
Venus	0.7	0.6
Earth	1.0	1.0
Mars	1.5	1.9
Jupiter	5.2	11.9
Saturn	9.5	29.5
Uranus	19.2	84.0
Neptune	30.2	164.8

QUESTIONS

1 Identify what represented the Sun and the planets in this model.
2 What was the relationship between distance from the tubing and time that it took the stopper to orbit?
3 What can you infer about the movement of the inner and outer planets from this model?
4 The table at the left contains data for the distance from the Sun and orbital times for each planet in our solar system. Create a line graph of this data.
5 Compare the trendline of your results with that of this second graph. How are they similar? How are they different?

CONCLUSION

Copy and complete:

'The results show that: (*respond to the aim*).'

Investigation **6.5A**

Making a telescope

KEY SKILL

WRITING A RESEARCH QUESTION

Turn the aim of this investigation into a question that asks what you are trying to discover. This is called a research question.

Hint #1: *Make sure that your research question has a question mark at the end.*

Hint #2: *Your research question can also be used as a title for an experiment report.*

AIM

To investigate how to use lenses to make a telescope

MATERIALS

- two convex lenses of different sizes
- ruler
- cardboard tube
- box cutter
- book, or sheet of paper with writing on it
- sticky tape
- Blu-Tack

METHOD

1. Hold the larger lens between you and the book or sheet of paper. The writing should look blurry when looking through the lens. This will be the primary or objective lens.
2. Hold the smaller lens between you and the larger lens. This will be the eyepiece lens. Move the lenses so that when you view the writing through the smaller lens it is in focus.
3. Record the distance between the first and second lenses.
4. Cut a slot in the cardboard tube about 2 cm from one end to hold the larger lens.
5. Cut a second slot for the smaller lens, the same distance away from the larger lens that you recorded in step 3.
6. Cut away any excess tubing, leaving about 2 cm.
7. Place the two lenses in the slots, holding them in place with tape or Blu-Tack.

QUESTIONS

1. Draw and label a scale diagram of your telescope. Include the objective lens, eyepiece lens and the measurements.
2. What is the purpose of the objective lens?
3. What is the purpose of the eyepiece lens?
4. What would you expect to happen with the detail you can see if you increase the size of these lenses? Explain.
5. What would you expect to happen with the length of the telescope if you increase the size of these lenses? Explain.

CONCLUSION

Copy and complete:

'The results show that: (*respond to the aim*).'

Investigation **6.5B**

Lens diameter and resolution

KEY SKILL

REPRESENTING DATA TO IDENTIFY PATTERNS AND TRENDS

When you write a formal investigation report, there is always a results section that includes your data, often as a table, chart or image. Choosing how to represent your data so that it can be clearly communicated to someone reading your investigation report is an important skill. In this investigation, after you have collected and recorded your data in the results table, turn your data into a chart or a graph. Identify any patterns or trends in your data and include this in your discussion.

Hint #1: *There are many ways to visualise your data, such as bar charts, line graphs and pie charts. Make sure you choose the best one for your data set.*

AIM

To investigate how lens diameter affects resolution

MATERIALS

- A4 piece of paper
- sticky tape
- marker pen
- ruler
- trundle wheel or long tape measure
- selection of binoculars and/or optical telescopes

RESULTS TABLE I6.5B

Optical instrument	Diameter of lens (mm)	Distance from 'stars' where they are first observed as two separate objects (m)
Human eye	10	
Binoculars		
Telescope		

METHOD

1. Copy the results table into your notebook, adding a title.
2. Measure and record the diameter of the lens of the binoculars and optical telescope and record it in the results table.
3. On the piece of paper, draw two thick lines 2 cm long and 2 cm apart. These will represent stars.
4. Use the sticky tape to tape the paper to a wall.
5. Position an observer at a point away from the wall where they observe the two lines as one.
6. The observer should walk towards the wall, stopping when they first observe the two lines as being distinct from one another.
7. Use the tape measure or trundle wheel to measure the distance between the observer and the wall. Record this in your results table.
8. Repeat steps 5–7 using the binoculars and telescopes that you have.
9. Construct a graph of your results.

QUESTIONS

1. What is the relationship between the diameter of the lens and the distance where the 'stars' are first observed as two separate objects?
2. Resolution is the ability to tell two separate objects apart, or the ability to see more fine detail. Which optical instrument had the best resolution and which had the worst?
3. Alpha Centauri is the brightest star in the constellation of the Pointers, and the third brightest star in the night sky. To the naked eye, Alpha Centauri appears as one star but it is a triple star system. Explain how resolution of an instrument is important for making these discoveries.
4. Compare the resolution of a telescope with the resolution of a microscope in the ability to distinguish between objects that are close together.

CONCLUSION

Copy and complete:

'The results show that: (*respond to the aim*).'

Investigation **7.1**

Galileo's pendulum

KEY SKILL

REPRESENTING AND RECORDING DATA USING A TABLE

In this investigation it is critical to record your data and observations using a suitable table. Before you start your investigation, design a table with appropriate columns and rows to record your data and observations.

Hint #1: *Use a ruler to draw your table so it is neat, clear and easy to read.*

Hint #2: *Give your table a title and a table number.*

Hint #3: *Include any units of measurement in the column or row headings.*

AIM

To investigate the law of conservation of energy by using a pendulum

MATERIALS

- metre ruler
- fishing line or poly string
- 50 g mass
- mass carrier (50 g)
- retort stand
- bosshead and clamp
- rubber stopper with hole
- pencil or pen

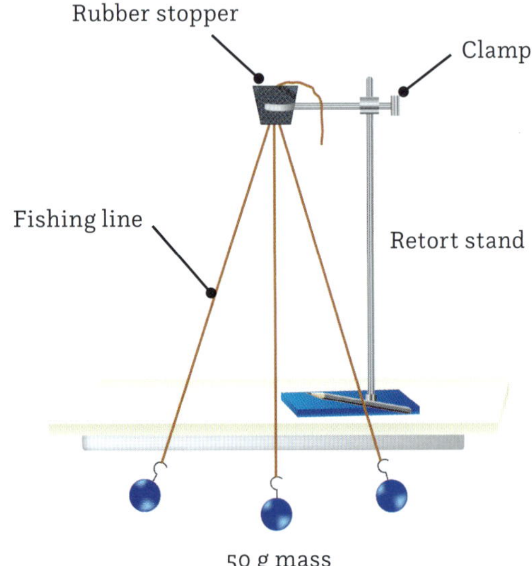

METHOD

1 Draw up a results table in your notebook.
2 Tie the mass carrier to the end of the fishing line, forming a pendulum.
3 Pass the other end of the fishing line through the rubber stopper and tie it off.
4 Insert the rubber stopper into a clamp on a retort stand as shown in the diagram.
5 Keeping the string taut, raise the mass up until the string is at about a 45° angle to the ground. Use the metre ruler to measure how far above the bench this is.
6 Release the mass, allowing it to swing backwards and forwards. Use the metre ruler to measure the height the mass returns to on its first swing. Record this height in your results table.
7 Interrupt the swing of the pendulum by placing a pencil in the path of the string. What height does the pendulum return to now?

QUESTIONS

1 Compare the return height to the starting height without the pencil in place. What did you notice?
2 Did the return height change when the pencil was placed in the path?
3 How could this experiment be improved?
4 How does this investigation demonstrate the conservation of energy?
5 Account for any energy losses in this investigation.

CONCLUSION

Copy and complete:

'The results show that: (*respond to the aim*).'

Investigation **7.2**

Energy efficiency of bouncing balls

KEY SKILL

IDENTIFYING THE VARIABLES AND FORMULATING A HYPOTHESIS

Before you formulate a hypothesis, identify your independent, dependent and controlled variables.
The independent variable is the one thing that you purposefully want to change in an investigation.
The dependent variable is what you will be measuring.
The controlled variables are all the things you need to keep the same throughout the investigation.
To formulate your hypothesis, use the following sentence stem: It can be hypothesised that if (something to do with your independent variable), then (something to do with your dependent variable).

Hint #1: *If you get stuck, use the prompts on page 127 to help you.*

Hint #2: *In your conclusion, state whether your hypothesis was supported (correct) or rejected (incorrect).*

AIM

To investigate the energy efficiency of different types of balls

MATERIALS

- metre ruler
- sticky tape
- several different types of balls; e.g. tennis ball, golf ball, squash ball, basketball, superball, hi-bounce ball

METHOD

1. Copy the results table into your notebook, adding a title and rows as needed.
2. Tape the metre ruler vertically to the side of a desk or bench, so that the zero mark is at the bottom.
3. Take the first type of ball and hold it so that the 1-metre mark is just visible under the ball.
4. Drop the ball, measuring the maximum height the bottom of the ball reaches as it bounces back up. Record your results in your results table.
5. Repeat step 4 for two more trials of the same ball.
6. Repeat steps 3–5 for each type of ball.
7. Calculate the efficiency (%) of each type of ball by dividing the average rebound height by the drop height.

QUESTIONS

1. Which balls were the most and the least energy efficient?
2. Which ball lost the most energy?
3. How could this experiment be improved?
4. Is there a link between the loudness of the bounce and the energy efficiency of the ball?
5. Name two possible sources of energy loss for a bouncing ball.

CONCLUSION

Copy and complete:
'The results show that: (*respond to the aim*).'

RESULTS

TABLE 17.2

Type of ball	Drop height (cm)	Rebound height				Efficiency
		Trial 1 (cm)	Trial 2 (cm)	Trial 3 (cm)	Average (cm)	$\frac{\text{average rebound height}}{\text{drop height}} \times 100$
Basketball	100					
Tennis ball	100					
	100					
	100					

Investigation **7.4**

Insulation and heat transfer

KEY SKILL

IDENTIFYING LIMITATIONS TO THE METHOD AND SUGGESTING IMPROVEMENTS

When you write a formal investigation report, there is always a discussion section that includes a discussion of potential errors. These errors are limitations (or problems) with the method. For each error, you list a way to control it (your suggested improvement).

Hint #1: *Brainstorm three potential errors that might have occurred in this investigation that could have affected or changed the results you collected (for example, if through human error something was not measured accurately). Now work with a partner to suggest ways each error could be controlled.*

AIM

To investigate the relationship between insulating material and the transfer of heat

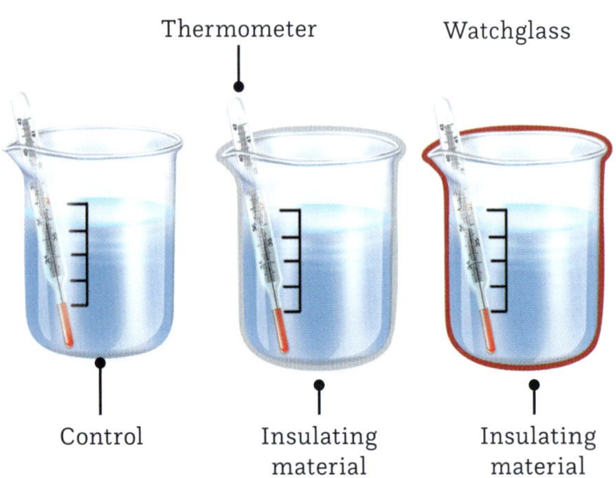

RESULTS

TABLE I7.4

Time (min)	Water temperature (°C)			
	Control	Material A	Material B	Material C
Initial				
5				
10				

MATERIALS

- selection of insulating materials, e.g. polystyrene, wool, aluminium
- 250 mL beakers (one for each material)
- thermometers (one for each beaker)
- watchglasses (one for each beaker)
- hot water
- scissors
- tape

METHOD

1. Copy the results table into your notebook, adding a title and rows as needed.
2. Wrap the insulating material around each beaker and secure with tape. Leave one beaker free of insulating material. This will be your control.
3. Create a lid for each beaker with the insulating material.
4. Add equal volumes of hot water to the beakers.
5. Place a thermometer in each beaker and secure the watchglass lid with tape.
6. Allow enough time for the temperature reading on the thermometer to stabilise. Record this initial temperature in your table.
7. Record the water temperature for each beaker every 5 minutes for half an hour.

QUESTIONS

1. Which material was the best insulator? Give evidence to support your answer.
2. Which material was the worst insulator? Give evidence to support your answer.
3. Choose the results from one material to graph.
4. Suggest a design for a thermos that aims to keep tea hot for as long as possible. What kind of materials could be used?

CONCLUSION

Copy and complete:

'The results show that: (*respond to the aim*).'

Investigation **7.5**

Atmospheric convection and the inversion layer

TEACHER DEMONSTRATION

AIM

To model how thermal pollution affects circulation in the atmosphere

MATERIALS

- 2 identical small, wide-mouthed jars
- stirring rod
- large, shallow tray
- hot water from a tap
- very cold water
- red food colouring
- blue food colouring
- cardboard or waxed paper

METHOD

PART 1:
INVESTIGATING CONVECTION

1 Your teacher will half-fill one of the jars with iced water. They will add 1 drop of blue food colouring. Observe how it mixes. Your teacher will mix the food colouring in completely if needed.
2 Your teacher will slowly add more iced water to the jar until there is a slight bulge of water over the top of the rim of the jar. Then they will set this jar in the shallow tray.
3 Your teacher will carefully half-fill the other jar with hot water and add 1 drop of red food colouring. Observe how it mixes. Your teacher will mix the food colouring in completely if needed.
4 Observe as your teacher fills this jar with hot water all the way to the top and places the card or waxed paper over it. They will tap the card gently so that the card forms a seal with the rim of the jar.
5 Your teacher will take the jar with hot water in one hand, with the other hand on the card. They will quickly, and without hesitating, pick it up and flip the jar upside down.
6 Your teacher will place the card and the jar of hot water on the jar of cold water so that the rims line up exactly.
7 Your teacher will carefully remove the card without disturbing the jars. Record your observations.

PART 2:
INVESTIGATING INVERSION

1 Your teacher will empty both jars and rinse them.
2 Watch as they repeat the steps in part 1, but this time put the jar with the red coloured hot water in the tray and put the card on top of the jar with the blue coloured cold water. They will turn the blue jar upside down and put it on top of the jar with red coloured hot water. Record your observations.

QUESTIONS

1 What happened in part 1 of the investigation? What happened in part 2, and how was it different?
2 Which jar represented the atmosphere with thermal pollution, and which represented the atmosphere with smog?
3 Why does the water mix so quickly when the glass of hot water is on the bottom?
4 If this situation happened in a large city, what would the air be like for the people living there?

CONCLUSION

Copy and complete:
'The results show that: (*respond to the aim*).'

Investigation **8.1**

Acceleration changes due to mass

KEY SKILL
REPRESENTING DATA TO IDENTIFY PATTERNS AND TRENDS

When you write a formal investigation report, there is always a results section that includes your data, often as a table, chart or image. Choosing how to represent your data so that it can be clearly communicated to someone reading your investigation report is an important skill. In this investigation, after you have collected and recorded your data in the results table, turn your data into a chart or a graph. Identify any patterns or trends in your data and include this in your discussion.
Hint #1: *There are many ways to visualise your data, such as bar charts, line graphs and pie charts. Make sure you choose the best one for your data set.*

AIM

To investigate the effect of increasing mass on an object's acceleration

MATERIALS

- dynamics trolley
- mass carrier
- 2 × 100 g masses
- 2 × 200 g masses
- 500 g mass
- bench mounted pulley
- fishing line
- stopwatch
- masking tape

METHOD

1 Copy the results table into your notebook, adding a title.
2 On a desk or benchtop, place two strips of masking tape about 1 metre apart. These strips represent the start and finish lines. Make sure there is at least a trolley's length of benchtop before and after each strip of tape.

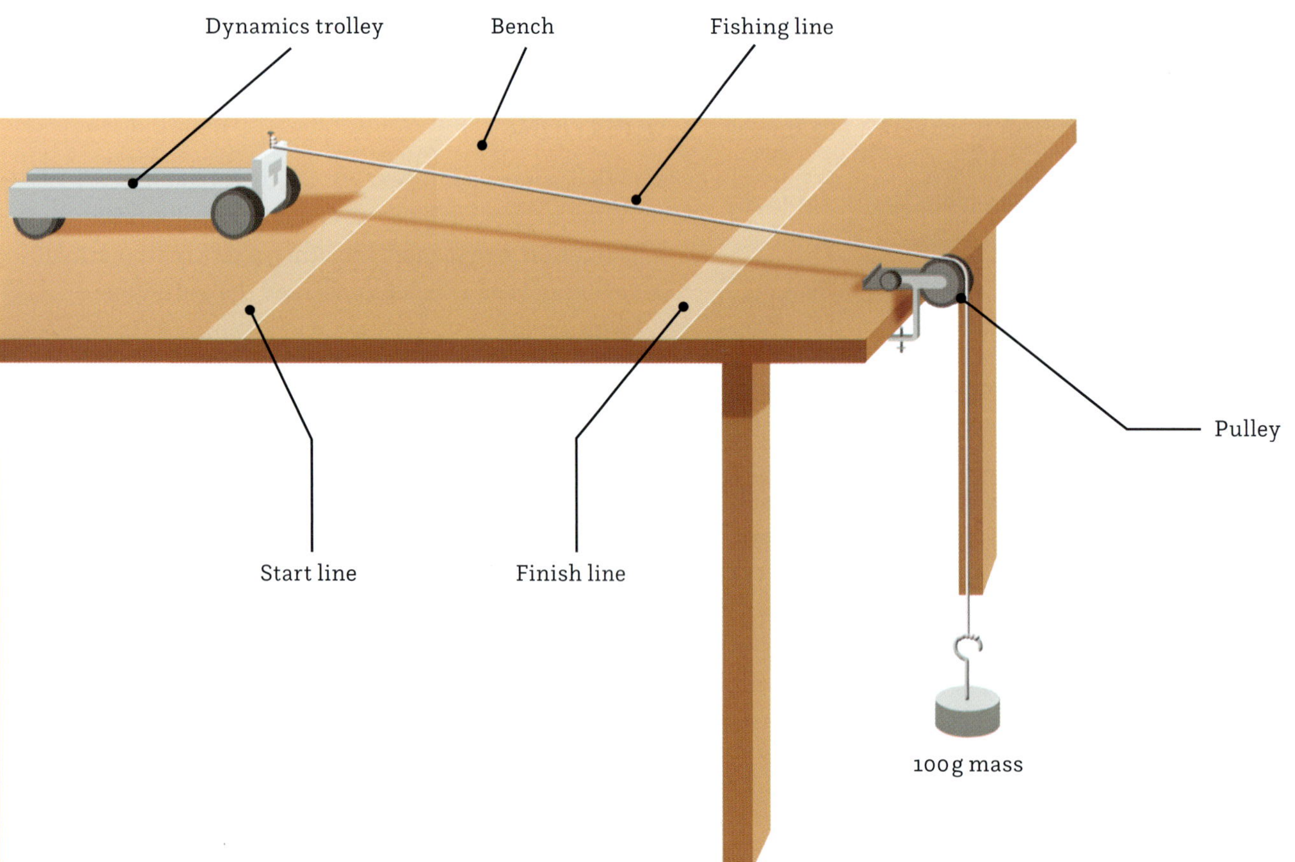

RESULTS **TABLE I8.1**

Mass on trolley (g)	Time (s) for trolley to travel from start line to finish line			
	Trial 1	Trial 2	Trial 3	Average
100				
200				
300				
400				
500				

3 Attach a 100 g mass to one end of the fishing line. Thread the other end over the pulley and attach it to the dynamics trolley, as shown in the diagram.
4 Release the mass so that it drops straight down to the ground. Use the stopwatch to time how long the trolley takes to travel from the start line to the finish line. Make sure to catch the trolley before it falls off the table. Perform three trials and record the times in your results table.
5 Add a 100 g mass to the trolley and repeat step 4.
6 Repeat step 4 with masses of 300, 400 and 500 g on the trolley. Record all your data.

QUESTIONS

1 Do your results follow a pattern? If so, describe what this is.
2 How do your results confirm Newton's second law?
3 What errors may have occurred in this experiment, and how could you overcome them next time?
4 Which part of the apparatus provides the force that accelerates the trolley?
5 Are there any other forces acting on the trolley? If so, what are they?

CONCLUSION

Copy and complete:
'The results show that: (*respond to the aim*).'

Investigation **8.2**

Ticker timers

KEY SKILL

IDENTIFYING LIMITATIONS TO THE METHOD AND SUGGESTING IMPROVEMENTS

When you write a formal investigation report, there is always a discussion section that includes a discussion of potential errors. These errors are limitations (or problems) with the method. For each error, you list a way to control it (your suggested improvement).

Hint #1: *Brainstorm three potential errors that might have occurred in this investigation that could have affected or changed the results you collected (for example, if through human error something was not measured accurately). Now work with a partner to suggest ways each error could be controlled.*

AIM

To investigate the relationship between speed, distance and time, using a ticker timer machine.

MATERIALS

- ticker timer
- 1 m strip of ticker tape
- 50 g mass
- sticky tape
- scissors
- power-pack
- 2 × electrical leads

METHOD

1 Copy the results table into your notebook, adding a title and rows as needed.
2 Use the sticky tape to attach the end of the ticker tape to the 50 g mass.
3 Thread the other end of the ticker tape through the ticker timer.
4 Start the ticker timer and drop (don't throw!) the 50 g mass to the ground.
5 Disconnect the tape from the mass and look at the tape. There should be a lot of ink dots on the tape. If the dots are difficult to see, repeat steps 2–4 with another strip of tape.
6 Draw a line through the first clear dot, and then every fifth dot after that. This represents a time of 0.1 seconds.
7 Measure the distance between each marked dot. This represents how far the mass travelled in each 0.1 second interval. Record the distance travelled in each interval in your results table. Note: You will need extra rows because you will have more than three intervals.
8 Find the speed of each interval in cm/s by dividing the distance travelled by the time taken (0.1 seconds). Enter the speed of each into the table.

RESULTS

TABLE I8.2

Interval	Time of interval (s)	Distance travelled (cm)	Average speed (cm/s)
1	0.1		
2	0.1		
3	0.1		

QUESTIONS

1 What do you notice about the speed of each interval? What does this tell you about the speed of the mass through its journey?
2 What errors were present in the experiment?
3 What does this experiment indicate about the speed of a falling object?
4 If you performed this experiment using a high-speed video camera instead of a ticker timer, do you think you would record the same speeds? Why or why not?

CONCLUSION

Copy and complete:
'The results show that: (*respond to the aim*).'

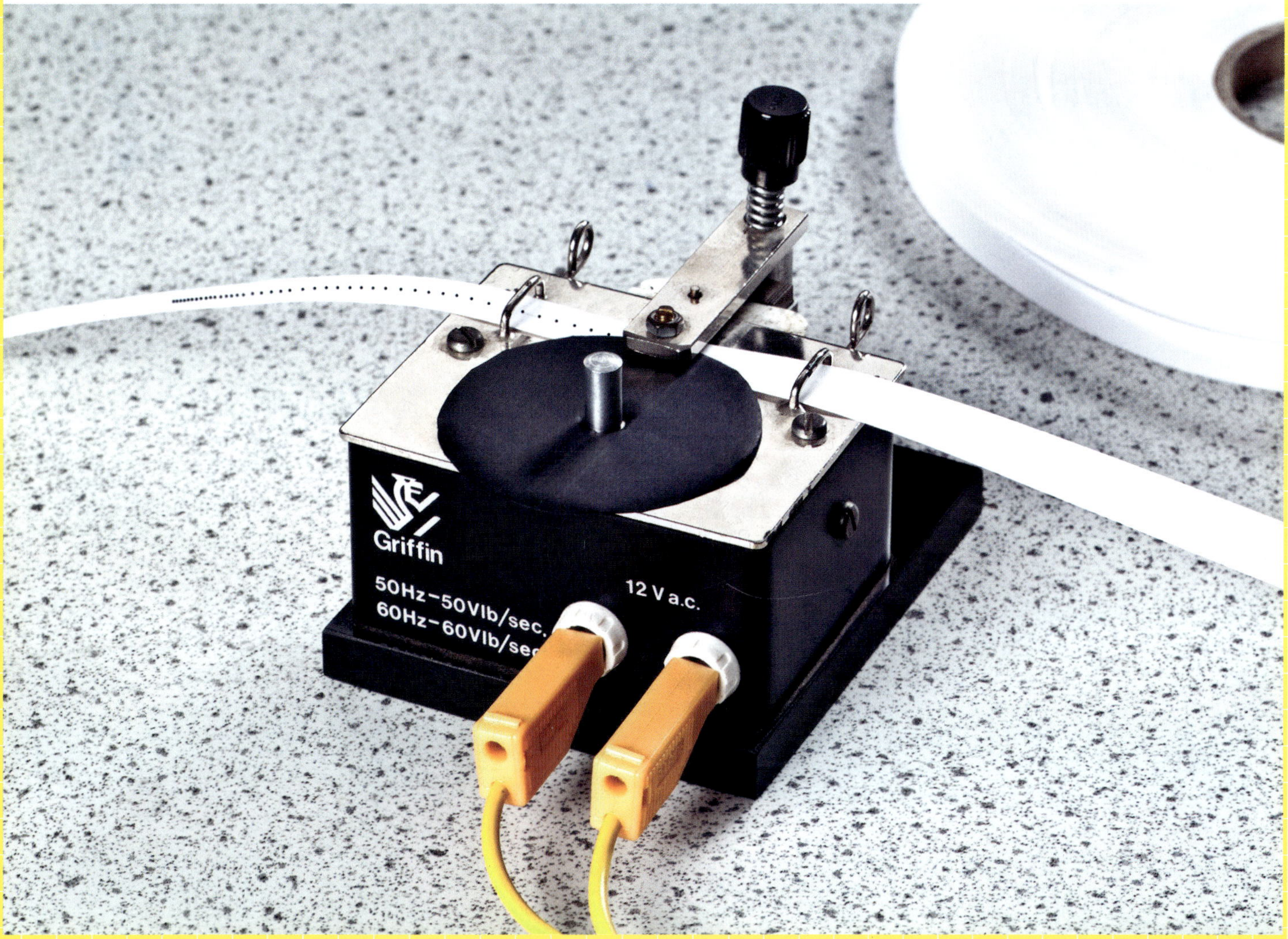

Investigation **8.3**

Vehicles and pedestrians

KEY SKILL

REPRESENTING AND RECORDING DATA USING A TABLE

In this investigation it is critical to record your data and observations using a suitable table. Before you start your investigation, design a table with appropriate columns and rows to record your data and observations.

Hint #1: *Use a ruler to draw your table so it is neat, clear and easy to read.*

Hint #2: *Give your table a title and a table number.*

Hint #3: *Include any units of measurement in the column or row headings.*

AIM

To investigate the speed travelled by vehicles and pedestrians

MATERIALS

- stop watch
- coloured markers or cones
- trundle wheel

METHOD

PART A

RECORDING THE SPEED OF VEHICLES

1. Draw up a results table for Part A in your notebook.
2. Use a trundle wheel to measure out 30 m along the footpath beside the road. Place cones or markers at each end of the 30 m space.
3. Position one student at the start and one at the end of the 30 m space.
4. One student uses a stop watch to time how long it takes for a car to travel from the starting marker to the finishing marker. Record this time in the results table.
5. Repeat step 4 for a total of 10 cars. If any pedestrians or cyclists travel the length of the 30 m space, record their time as well in order to compare it later to the speed the cars are travelling.

PART B

RECORDING THE SPEED OF PEDESTRIANS

1. Draw up a results table for Part B in your notebook.
2. Use the trundle wheel to measure out 10 m along the footpath beside the road. Place cones or markers at each end of the 10 m space.
3. Position one student at the start and one at the end of the 10 m space.
4. One student uses a stop watch to time how long it takes for Student 1 to walk the 10 m space. Record this time in the results table.
5. Repeat step 5, but instead of walking normally, record the speed for walking backwards, hopping, running and an activity of the student's choosing.
6. Students switch places and collect data for Student 2.

QUESTIONS

1. What is the formula used to calculate speed?
2. Consider the speed limit of the road you were on. Were any vehicles over the speed limit?
3. Which activity had the fastest student speed? Which had the lowest?
4. What errors may have led to inaccurate results during this investigation? How could you improve the method to ensure better accuracy?

CONCLUSION

Copy and complete:

'The results show that: (*respond to the aim*).'

Investigation **8.4**

Balloon rockets

KEY SKILL
IDENTIFYING AND MANAGING RELEVANT RISKS

Brainstorm with a partner to identify three possible hazards or risks that may be involved in this investigation. Suggest one way that each hazard or risk could be reduced.

AIM

To demonstrate Newton's third law, using balloon rockets

MATERIALS

- balloon
- fishing line
- sticky tape
- large drinking straw

METHOD

1 Inflate the balloon to about the size of a basketball. Hold the mouth of the balloon closed, but don't tie it off.
2 Attach the straw to the balloon as shown in the diagram. Thread the fishing line through the straw.
3 Ask a fellow student to hold one end of the fishing line while you hold the other, keeping it taut. Make sure the balloon is near one end of the fishing line so that most of the line is out the front of the balloon rocket.
4 Let go of the mouth of the balloon. Record your observations.

QUESTIONS

1 Describe the motion of the balloon relative to the air escaping from it.
2 How did the motion of the balloon demonstrate Newton's third law of motion? (Hint: In which direction did the balloon push the air?)
3 How could the set-up be improved to allow the balloon to travel further and faster?
4 What provided the force that accelerated the balloon?
5 If you inflate a balloon and then let it go when it isn't attached to a straw or the line, it flies all over the place in an unpredictable path. Why do you think this is?

CONCLUSION

Copy and complete:
'The results show that: (*respond to the aim*).'

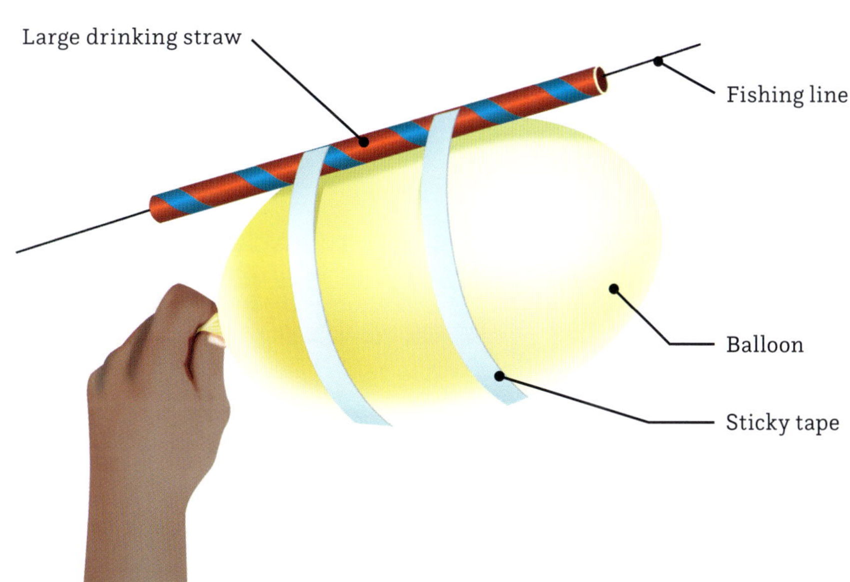

Investigation **8.5**

Car safety

KEY SKILL

IDENTIFYING THE VARIABLES AND FORMULATING A HYPOTHESIS

Before you formulate a hypothesis, identify your independent, dependent and controlled variables.

The independent variable is the one thing that you purposefully want to change in an investigation.

The dependent variable is what you will be measuring.

The controlled variables are all the things you need to keep the same throughout the investigation.

To formulate your hypothesis, use the following sentence stem: It can be hypothesised that if (something to do with your independent variable), then (something to do with your dependent variable).

Hint #1: *If you get stuck, use the prompts on page 127 to help you.*

Hint #2: *In your conclusion, state whether your hypothesis was supported (correct) or rejected (incorrect).*

AIM

To investigate Newton's laws by using a model of a crash test dummy

MATERIALS

- plasticine
- dynamics trolley
- talcum powder
- ramp
- metre ruler
- brick

METHOD

1 Copy the results table into your notebook, adding a title.

2 Make three plasticine models – small, medium and large. These will represent a baby, a child and an adult.

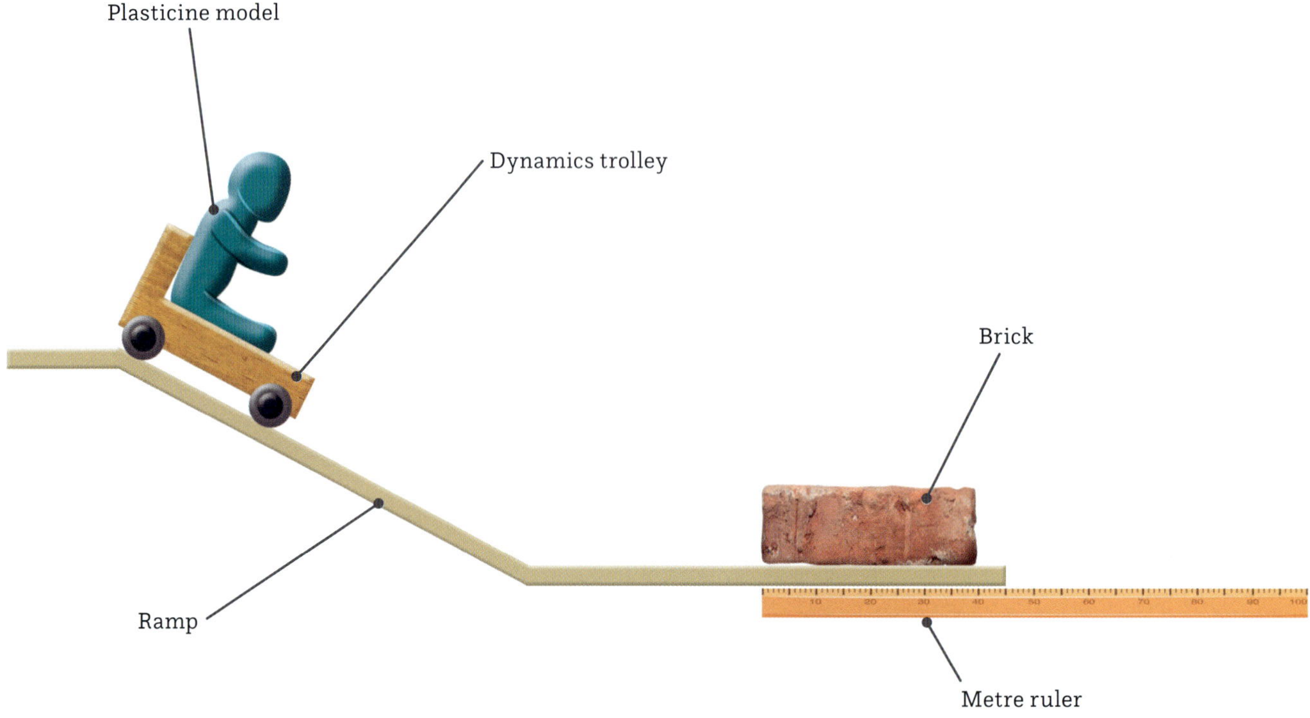

3 Lightly dust the top of the dynamics trolley with talcum powder, and sit the 'baby' on top. Make sure it isn't stuck down.
4 Set up the ramp and the brick as shown in the diagram. Place the metre ruler next to the brick so that the 0 cm mark is at the near edge of the brick.
5 Place the trolley with the plasticine 'baby' at the top of the ramp and then release it. The trolley should travel down the ramp and strike the brick at the bottom.
6 Record how far the 'baby' travels after striking the brick.
7 Repeat steps 5 and 6 twice to obtain two more results. Calculate the average distance from the three trials.
8 Repeat steps 5–7 for the 'child' and 'adult' models

QUESTIONS

1 On average, which sized person travelled the farthest after striking the brick? Why do you think this is?
2 How could this experiment be improved to give more accurate results?
3 How does this experiment show that it is important to always wear a seatbelt in a car?
4 Describe how each of Newton's three laws can be demonstrated from one aspect of this experiment. (Hint: What happened to the motion of the brick and the cart when they collided?)

CONCLUSION

Copy and complete:
'The results show that: (*respond to the aim*).'

RESULTS TABLE I8.5

Plasticine model	Distance travelled (cm)			
	Trial 1	Trial 2	Trial 3	Average
Baby model				
Child model				
Adult model				

GLOSSARY

acceleration a change in speed or direction of motion over time

accuracy how close the results are to the true values

action–reaction pair two equal and opposite forces exerted by two objects on each other

activation energy the energy required to start a reaction

adaptation a change in the structure or function of an organism, evolved over generations, that makes it better suited to its environment

allele a different version of a gene

antimatter particles that have properties opposite to that of normal matter

astronomer a scientist who studies space, stars and celestial objects

astronomical unit the average distance between Earth and the Sun (about 150 000 000 km)

astrophysicist a scientist who studies the physics of the universe

atmosphere the layer of gas that surrounds Earth

atom the smallest particle of matter, made up of electrons orbiting a nucleus of protons and neutrons

atomic number the number of protons in an atom

balanced force a force acting on an object that is cancelled out by another force, so the net force is zero

biodiversity the variety of living species on Earth

biogeography the study of the past and present distribution of living organisms

biosphere all of the living things on Earth

biotechnology the use of living organisms to develop products

Bohr diagram a diagram showing the electron configuration of an atom

carbon capture the process of trapping carbon dioxide at its emission source so that it does not enter the atmosphere

carbon cycle the cycle that explains how carbon moves between Earth's spheres

carbon sink a place where carbon is stored in the carbon cycle

catalyst substance that increases the rate of a chemical reaction without being used up

cellular respiration the process that all living things use to produce cellular energy from glucose and oxygen

coefficient a number placed before the chemical in a formula or chemical equation

chemical formula chemical symbols showing the ratio of elements to one another

chemical property a property of a substance that is observed during a chemical reaction

chemical symbol one or two letters that represent an element

chromosome a tightly coiled strand of DNA

circulate to move freely or continuously through an area

comparative anatomy the study of features of different species to look for evidence of a common ancestor

compound a substance made up of two or more types of atoms bonded together in fixed ratios

concentration the number of particles in a given volume

conservation of mass the law that mass cannot be created or destroyed

controlled variables all the things that need to stay the same during an investigation

covalent bond a bond in which two atoms share one or more pairs of electrons

Dalton's atomic theory the theory that states that all matter is made up of tiny particles

dependent variable the thing that will be measured and is altered by the independent variable

deoxyribonucleic acid (DNA) found in the nucleus of a cell; the carrier of genetic information

displacement the distance an object is from its starting position

distance the total length an object travels

double helix the structure of a DNA molecule; a double-stranded spiral

ductile can be drawn out into a wire

efficient not wasteful

electron configuration the organisation of electrons into shells around the nucleus

electron shells orbits around the nucleus that hold electrons of similar energy levels

element a pure substance made up of one type of atom

energy a measure of the ability to do work

energy efficiency how much usable energy is produced compared to how much energy has been supplied

energy transfer the movement of energy from one place to another without changing form

energy transformation a change from one type of energy to another

enhanced greenhouse effect an increase in the greenhouse effect due to human greenhouse gas emissions

enzyme a biological catalyst that increases the rate of reactions in cells

error the difference between the measured value and the actual value

ethical consideration a factor that takes into account what is right and what is wrong

evolution the gradual change in heritable characteristics over time

exoplanet a planet outside our solar system

experiment an investigation carried out under controlled conditions, to test a hypothesis

fair test an investigation in which only one factor is changed and all other variables are kept the same

fertilisation the fusion of male and female gametes to form a zygote

fieldwork an investigation conducted in the natural environment, not a laboratory

finite limited; won't last forever

force a push, pull or twist

fossil the geologically altered remains of an organism that was once alive

fossil fuel a fuel that is formed from the decomposition of dead animals and plants over millions of years; e.g. oil and coal

fossil record a record of all fossils found on Earth

friction a contact force that opposes motion, caused by objects rubbing against each other

frost line a boundary just inside Jupiter's orbit

galaxy a system of millions or billions of stars

gamete a sex cell – an ovum or a sperm

gene a segment of DNA, the basic functional unit of heredity

generate to produce or make something

genetic variation the differences in genes within individuals of a population

genetically modified (GM) an organism that has had alterations to its DNA

genome an organism's entire set of DNA

genotype the genetic code for a gene or an organism's entire genome

greenhouse effect the trapping of the Sun's warmth by the atmosphere

greenhouse gas a gas that traps heat energy in the atmosphere

group a column in the periodic table

heredity the passing of traits from parents to their offspring

heritable a characteristic that can be passed down from parent to offspring

heterozygote a gene that has two different alleles present

homologous structures parts of organisms that show evidence of a common ancestor

homozygote a gene that has identical alleles present

hydrosphere all the water on Earth

hypothesis a scientific statement that can be tested

implication a likely outcome or consequence of an action

in vitro fertilisation (IVF) fertilisation of the sperm and egg that occurs outside the human body

independent variable the thing that is purposefully changed during an investigation

inertia a property of matter that causes it to resist change in speed or direction (to remain at rest or in a state of uniform motion)

inference an educated guess or judgement based on observations

insulation material that acts as a barrier to heat flow

inversely proportional when one value decreases at the same rate that the other increases

ion an atom with unequal numbers of protons and electrons; a charged atom

ionic bond a bond in which some atoms gain and some lose electrons, becoming ions

isolation the different adaptations of members of a species that have been separated from the rest of the population

isotope an atom of a particular element with a different number of neutrons

joule the unit of energy

karyotype a picture of an organism's full set of chromosomes

kinetic energy energy of motion

lattice an interlaced structure or pattern

light-year the distance that light travels in one Earth year

lithosphere Earth's rigid outer zone (crust and upper mantle), made up of tectonic plates

malleable can be bent

mass the amount of matter in an object

mass number the sum of the protons and neutrons in an atom

matter a physical substance that has mass and takes up space

mean the values of a data set added together and divided by the number of values

median the middle result in a set of data

meiosis complex cell division that produces gametes (sex cells)

metallic bond a bond in which free electrons move around metal ions

metalloid an element with properties similar to both metals and non-metals

mistake something that has been done incorrectly in an experiment

molecule the smallest unit of a covalent compound that can take part in a chemical reaction

mutation an error in DNA caused by a substitution, an insertion or a deletion of a base

natural selection the process whereby organisms better suited to their environment tend to survive and reproduce, passing on traits to future generations

nebula a vast region of gas and dust

net force all the forces acting on an object added together

neutron star an extremely dense star left over after a supernova

newton the unit of measurement of force

normal force the force of the floor or ground pushing an object upwards

nucleotide building block of DNA (adenine, guanine, thymine or cytosine)

observation something you see and know to be true

ovum a female sex cell, also known as an egg

ozone a molecule made of three oxygen atoms bonded together

ozone-depleting substance a chemical that reacts with ozone and breaks the molecule down as well as stopping the molecule from re-forming

ozone layer a region of the stratosphere with a high concentration of ozone

palaeontologist a scientist who studies fossils

parsec 3.26 light-years

passive design building design that considers climate and location to make the best use of natural heating and cooling

period a row in the periodic table

permafrost frozen soil in the Arctic

phenotype how a genotype is physically expressed

photosynthesis the process that plants use to make glucose from carbon dioxide and water

physical property a property of a substance that can be measured, such as colour, melting and boiling points, density and hardness

polyatomic ion two or more ions bonded together and acting as a single charged unit

precision how close the results of repeated trials are to each other

prediction a statement about the future based on observation and evidence

predisposition the increased likelihood of developing a disease or condition

prefix a word or number placed before another word

primary data first-hand data, from your own investigation

product a substance formed in a chemical reaction

pure substance a substance that contains only one type of particle

qualitative written descriptions and observations

quantitative numerical information and data

range the largest value in a data set minus the smallest value

reactant a substance that starts a chemical reaction

red giant a star that has stopped fusing hydrogen in its core

redshift a change in light's wavelength towards the red end of the visible spectrum

reliable provides consistent results when repeated

research to gather data and information in an organised way to inform a hypothesis or an investigation

resolution the ability to tell two separate objects apart

ribonucleic acid (RNA) a nucleic acid that carries instructions from DNA

sample size the number of participants in a survey or samples tested in an experiment

scalar a quantity that has a magnitude but no direction

secondary data second-hand data, from someone else

sensationalised language that aims to shock or produce an emotional reaction in the reader

singularity an infinitely dense point of matter that existed before the big bang

smog a form of severe air pollution

social consideration a factor that affects society and the people within society

speciation the creation of new and unique species

speculative relying on theories or opinions rather than facts

speed the distance an object travels divided by the time taken to travel that distance

sperm a male sex cell

stationary not moving; still

stem cell a cell type that is unspecialised and can develop into other types of cells

supernova an explosion of a massive star at the end of its life

surface area the area of the outermost layer of an object

surrogacy an arrangement in which a woman becomes pregnant on behalf of another person

temperature inversion weather conditions that stop fresh air from circulating

thermal to do with heat

trend a consistent change in the values of a data set

ultraviolet radiation harmful rays from the Sun, which can cause cancers and harm animals and plants

unbalanced force a force acting on an object that is not cancelled out by another force, so a net force acts on the object

uncertainty the margin of error of a measurement

usable energy the type of energy that a device is designed to produce

valence electrons electrons orbiting in the outer shell of an atom

valence shell the outermost shell of an atom

valency an atom's power to combine with other atoms

valid measures what is intended to be measured

variation the natural differences between individuals in a population

vector a quantity that has both a magnitude and a direction

velocity a measure of how quickly displacement changes

vestigial structures parts of organisms that have lost some or all of their original function

weight force the force generated by gravity pulling an object downwards

white dwarf a small, very dense star formed at the end of a small star's lifetime

INDEX

ACKNOWLEDGEMENTS

The author and publisher are grateful to the following for permission to reproduce copyright material:

PHOTOGRAPHS: AAP Photos/Luke Aikins, **111** (top), /AP Photo/ Antelope Valley Press, Ron Siddle, **14** (bottom), /Matt Dunham/ AP, **3** (top); Alamy/Bob Kreisel, **41**, **46** (neon sign), /Custom Life Science Images, **100** (bottom), /dpa picture alliance, **113** (bottom), /Historic Images / Alamy, **46** (Mosley), /Sebastian Kaulitzki, **14** (top), /Ivan Kuzmin, **26** (top), /Patti McConville, **133**, /PJF Military Collection, **48** (explosion), /Prisma by Dukas Presseagentur GmbH, **22** (left), /RGB Ventures/SuperStock, **85** (bottom), /Scenics & Science, **vi–1** (banner), /Science History Images, **20**, **iv** (bottom), **46** (Dobereiner), /Science Photo Libary, **44**, **46** (Newlands), **32** (potassium exploding), **179**, / David Shale/NPL, **24** (top), /Martin Shields, **127**; Douglas Mark Black Photographer/ Farm House by Archterra Architects, **105**; CERN/Anna Pantelia, **51** (top left); CSIRO/ David McClenaghan, **93**; DVIDS/ Shannon E. Renfroe/U.S. Navy, **114**; European Chemical Society (CC BY ND), **33** (top); ESA/Hubble/NASA, ESA, and S. Beckwith (STScI) and the HUDF Team, **83** (top), /ATG medialab, CC BY-SA 3.0 IGO, **91**; ESO/H. Drass et al, **86** (bottom), /M. Kornmesser, **83** (bottom), **94** (black hole); Event Horizon Telescope Collaboration, **87**; Flickr/ Lennart Halling, 1960-11 CC BY 2.0, **33** (bottom); Getty Images /Ulrich Baumgarten, **66** (woman), /BEDI/AFP, **49** (bottom), /Monica Bertolazzi, **88**, **94** (milky way), /Ben Curtis/PA Images/Getty Images, **12** (top), **16** (sheep), /C12, **71**, /Chase Dekker Wild-Life Images, **67** (top), /Jason Edwards, **21** (bottom), /Daniel Eskridge/Stocktrek Images, **19** (bottom), /Icon Sportswire, **112** (bottom), /Marka, **96** (crash test), /Cameron Spencer, **116**, /Sarah Silbiger / Stringer, **67** (bottom), /Schroptschop, **73**, /The Washington Post, **78** (top), /William West/AFP, **77**; iStock/:f3rd14n, **102**, /Bedrin-Alexander, **32** (atom), /Maximilian Butek, **iv–v** (banner), /Ekaterina79, **132**, /joingate, **129** (bottom left), /upthebanner, **46** (statue of libery), /1xpert, **66** (earth), /3dalia, **46** (iron element), /abzerit, **18** (chimp), /alanphillips, **9** (bottom), /angel_nt, **62**, **64** (power plant), /antpkr, **141** (pliers), /artursfoto, **163**, /belchonock, **48** (coffee), /benjamint444, **25**, /Betka82, **141** (crucible tongs), / bjdlzx, **82** (milky way), /bonchan, **63**, /BookyBuggy, **141** (kitchen tongs), /Boris25, **38**, /bymuratdeniz, **61**, /CarolinaSmith, **82** (top), /chaofann, **48** (molecular structure), /CoffeeAndMilk, **137**, /Cylonphtot, **i**, /Delpixart, **96** (wind turbines), /Design Cells, **27**, /Explora_2005, **78–9**, /fatihhoca, **110** (top), /frentusha, **143**, /GeorgiosArt, **122** (Isaac Newton), /GlobalP, **18** (top), **22** (right), /goir **113** (top left), /gokhanilgaz, **72** (orange coral), /graphicassault, **48**, **64** (rusted balde), /Gwenvidig, **66** (top), /hudiemm, **35**, /ImageMediaGroup, **99**, /imaginima, **122** (Newton's cradle), /jacoblund, **110** (boxer), /janiecbros, **2** (scientist), /JensenChua, **18** (kangaroos), /jeridu, **141** (forceps), /jlazouphoto, **72** (centre), /joggiebotma, **120** (top), /johan63, **82** (satellite), /JohnnyGreig, **129** (bottom right), / kali9, **126**, /Kanawa_Studio, **58** (left), /Lemonan, **58** (right), / lovleah, **18** (desert), /LSOphoto, **2** (twins), /lusia83, **108** (coal), / mehmettorlak, **72** (purple coral), /MicroStockHub, **100** (top), / Krystian Nawrocki, **108** (smartphone), /NeoPhoto, **156**, /Nerthuz, **86** (top), /NiseriN, **120** (bottom), /oversnap, **82** (eclipse), / paulafrench, **30** (African animals), /peepo, **110** (sprinter), / PeopleImages, **110** (man pushing car), /pjmorley, **80** (earth top), /Pongasn68, **167** (top), /Priyono, **66** (volcano), /rentusha, **12** (bottom), **16** (bottom), /rwarnick, **169**, /s-c-s, **30** (house construction), /Sergey_Peterman, **151**, /sharply_done, **96** (top), /Sjo, **97** (top), /skodonnell, **48**, **64** (shiny blade), /skynesher, **8**, /smirkdingo, **2** (cells), /so_illustrator, **21** (top), /South_ agency, **121** (bottom), /spxChrome, **48** (slightly rusted blade), /stevecoleimages, **32** (periodic table), /sweetmoments, **119**, **122** (motorcycle), /t:Grafner, **103**, /Tatomm, **48** (cleaning products), /ThamKC, **54**, **96** (smartphones), /Tsekhmister, **177**, /ultramarinfoto, **51** (top right), /vgajic, **98**, /VIDOK, **108** (flower), /wabeno, **113** (top right), /Максим Шмаков, **96** (power station), **108** (power plant), /Михаил Руденко, **110** (astronaut); NASA, **82** (gravitational pull), **94** (Hubble space telescope), **118** (bottom), /BARREL, David Milling, **75**, /ESA and the Hubble Heritage Team (STScI/AURA) – Hubble/Europe Collaboration, H. Bond (STScI and Pennsylvania State University), **111** (bottom), /WMAP Science Team, **85** (top), **94** (cosmic microwave background radiation); Nature Picture Library/ Dave Watts, **24** (bottom), **30** (Tasmanian devil); Science Photo Library, **45** (bottom), **97** (bottom), /TONY CAMACHO, **30** (hands), /Martyn F. Chillmaid, **55**, /Steve Gschmeissner, **19** (top), /ROB JUDGES / OXFORD UNIVERSITY IMAGES, **15**, /SCIEPRO, **19** (top left); Science Source/Francis Leroy, Biocosmos, **4**, /SCIENCE PHOTO LIBRARY, **42**, /Turtle Rock Scientific, **59**, /Charles D. Winters, **49** (top); Shutterstock/3Dalia, **34** (right), /blurAZ, **115**, /Darrel Camden-Smith, **46** (molten gold), /Virojt Changyencham, **64** (chalk board), /Chesky, **104**, /cigdem, **148**, /concept w, **36–7**, /Dotted Yeti, **89** (top), /sonya etchison, **2** (top), /Everett Historical, **34** (left), /kai hecker, **155**, /iviewfinder, **46** (motorcycle engine), / Kateryna Kon, **11**, **16** (karyotype), /Jatupol Leaupathanasuk, **112** (top), /Maximusmeridi, **18** (fossils), /milanzeremski, **122** (woman hitting tennis ball), /MrGoSlow, **56**, /Nerthuz, **94** (hydrogen), **86** (top), /osmandarinas, **43**, /Elena Pavlovich, **16** (stem cells), /posteriori, **66** (iceberg), /Rakic, **32** (top), /Vadim Sadovski, **94** (astronaut), /Yurchanka Siarhei, **2** (egg&sperm), /Tracy Starr, **140**, /stockphoto-graf, **161**, /theLIMEs, **170**, /Bodnar Taras, **118** (centre), /vchal, **46** (periodic table), /vectortatu, **89**, **94** (galaxies), /YuRi Photolife, **28** (top), **30** (fossil), /Marc Ward, **101**, /Steve Wilkins, **92**; Wellcome Collection (CCBY 4.0), **44**, **45**, **46** (Dalton and Mendeleev).

OTHER MATERIAL: The Victorian Curriculum F-10 content elements are © VCAA, reproduced by permission. The VCAA does not endorse or make any warranties regarding this resource. The Victorian Curriculum F-10 and related content can be accessed directly at the VCAA website, https://victoriancurriculum.vcaa.vic.edu.au/, **vi–1**.

While every care has been taken to trace and acknowledge copyright, the publisher tenders their apologies for any accidental infringement where copyright has proved untraceable. They would be pleased to come to a suitable arrangement with the rightful owner in each case.